Exchange and Production Systems in Californian Prehistory

The Results of Hydration Dating and Chemical Characterization of Obsidian Sources

Jonathon E. Ericson

BAR International Series 110
1981

B.A.R.

B.A.R., 122 Banbury Road, Oxford OX2 7BP, England

GENERAL EDITORS

A. R. Hands, B.Sc., M.A., D.Phil.
D. R. Walker, M.A.

B.A.R. S110, 1981: "Exchange and Production Systems in Californian Prehistory"

© Jonathon E. Ericson, 1981

The author's moral rights under the 1988 UK Copyright,
Designs and Patents Act are hereby expressly asserted.

All rights reserved. No part of this work may be copied, reproduced, stored, sold, distributed, scanned, saved in any form of digital format or transmitted in any form digitally, without the written permission of the Publisher.

ISBN 9780860541295 paperback
ISBN 9781407325071 e-book
DOI https://doi.org/10.30861/9780860541295
A catalogue record for this book is available from the British Library
This book is available at www.barpublishing.com

Dedicated to Cheryl Burke
and
The California Indians

PREFACE

The basis of this monograph is a dissertation that I submitted to the Department of Anthropology, University of California, Los Angeles, in late December 1977. This work has been edited and revised slightly. A calculation error was found in the section on induced hydration rates. The activation energies have been recalculated. In this version, the experimental and empirical rates are much more in agreement. Due to prior commitments on publication, two sections have been deleted. The long half-life neutron activation analyses will be published with Jerome Kimberlin. The other section is the UCLA radiocarbon date list which will be published with Professor Rainer Berger in a forthcoming issue of <u>Radiocarbon</u>. These two projects which involved considerable time and effort on the part of these researchers have been deleted without serious loss to the integrity of the larger work.

This monograph represents an attempt to understand the basis of development of regional exchange systems using an archaeometric approach. Since regional exchange systems are just beginning to be investigated (Earle and Ericson 1977), the focus of this study has been to develop new methods for describing and isolating systemic variables which control regional exchange. To do this the researcher must be able to determine both the chronology of the systems and the sources or origins of the materials under study. Obsidian lends itself to both dating through hydration dating and tracing through chemical characterization analysis. Research on obsidian as a material provides the necessary and basic data to undertake the analysis of production and exchange. For example, the analysis of quarry production in time has relied upon the obsidian hydration dating technique. For these reasons, the monograph is divided into two parts.

The first part of the monograph relies upon geophysical and geochemical methods in investigating basic questions about the obsidian hydration phenomenon as well as effectiveness of chemical characterization as a means to identify the origins of obsidian artifacts. The results of these preliminary investigations are used to define individual exchange systems and to describe the chronology of undiagnostic flakes and other production debitage on obsidian quarry workshops.

The second part of this work addresses theoretical questions regarding the evolution and maintenance of regional exchange systems in prehistoric California. Synagraphic mapping, mathematical modeling, obsidian hydration dating, chemical characterization, and quarry production analysis are some of the new techniques which are developed and tested in the course of this investigation.

Both major parts of this monograph follow a very similar strategy in addressing the respective research questions by examining a set of theoretical issues to propose a number of potential variables for analysis, by selecting and using a suite of scientific techniques for measurement of specific variables, and finally by analyzing the parameters which control and affect the systems under investigation.

ACKNOWLEDGEMENTS

Research which adds to the collective knowledge of mankind is the result of many individuals. We all share in this knowledge, drawing and adding as we are in need or capable. This monograph, which was a daily exercise for eight years, is no exception and is the result of the effort of many people, both directly and indirectly. I want to thank everyone for their assistance in this collaborative project.

The research on mechanisms of obsidian hydration dating was begun in May of 1969 under the guidance of Professor Rainer Berger who accepted me into his laboratory and then underwrote and supported much of the research herein. To him I extend my deepest thanks and remain indebted.

As an overview of the projects, research has been conducted under the auspices, guidance, and support of the following UCLA professors. Professor Rainer Berger, my dissertation committee chairman, was my advisor in almost all aspects of research. Through the Isotope Laboratory, Institute of Geophysics and Planetary Physics, UCLA, I was able to determine the many radiocarbon dates necessary for temporal control of the data. Professor Wayne Dollase, Department of Geology, UCLA, provided access to X-ray fluorescence instrumentation for analysis of the major elementary composition of obsidian samples. He has personally conducted Mössbauer experiments to determine the ferric-ferrous iron ratios in several obsidian samples. In addition, by his arrangement I have had the opportunity to present my early findings to the students and faculty of the Department of Geology, UCLA. Professor Timothy Earle, my Cochairman, Department of Anthropology, UCLA, encouraged and guided my research in prehistoric exchange system analysis for three years and, under his tutorage, I gained valuable anthropological perspective into the problems of prehistoric exchange. Professor John MacKenzie, Materials Department, School of Engineering and Applied Science, UCLA, was extremely helpful to me in gaining a theoretical understanding of obsidian as a glass. As my employer and advisor for a period of two years, he enabled me to analyze and compile most of the intrinsic properties of the obsidian samples present in chapter 3. His specific interests in obsidian as a natural glass and the hydration phenomena were very valuable in this undertaking. Professor Clement Meighan, Department of Anthropology, UCLA, as one of my initial sponsors, enabled me to understand the importance of obsidian hydration dating in the archaeology of Western North America. Moreover, through his guidance I was able to make the difficult transition from being solely a student of geophysics to the field of anthropology/archaeology. His support and counsel in undertaking the obsidian hydration research project and other overall aspects of archaeological research have been invaluable. Professor Dwight Read, Department of Anthropology, UCLA, helped me for three years to begin to master mathematical and statistical analysis which were used to analyze data and verify the proposed hypotheses. In addition, his counsel in many phases of research was invaluable.

I would like to acknowledge additional non-committee members for their support and contributions in this undertaking. Doctor Phillip Lankford, Department of Geography, offered me constant encouragement and guidance in undertaking prehistoric exchange system analysis through spatial analyses while he was at UCLA. Professor Howard Reiss, Department of Chemistry, UCLA, helped me to model the hydration phenomena as a complex thermodynamic process. Professor Thomas A. Tombrello, Department of Nuclear Physics, Cal Tech, conducted experiments using a linear accelerator to measure both the saturated and intrinsic water concentration values in several hundred obsidian samples. His interest in and support of this project are gratefully acknowledged. Professor John Wasson, Department of Chemistry and Geochemistry, and Institute of Geophysics and Planetary Physics, UCLA, supported most of my chemical characterization research through instrumental neutron activation analysis for over four years. Without his continued support, this project would not have been possible.

A special acknowledgement goes to my friend and colleague, Clay A. Singer, who perhaps was the most important in the difficult transition from geophysics to archaeology. With his patience and constant advice throughout these years, the transition was made easier. From the first Chapalan adventure to as recently as the thunderbolts at Bodie Hills, I look forward to years of collaborative effort with him.

I am deeply grateful to my friend and colleague, Jerry Kimberlin, who introduced me to the potential of neutron activation analysis in the early 1970s and who completed so many of the analyses himself. I look forward to more moderate collaborative projects than this one in the future. Also, I am indebted to Victoria Bennett of the UCLA Obsidian Hydration Laboratory who prepared and carefully read hundreds of obsidian hydration slides, even down to the last few days. Without her assistance, much of this work would not have been possible. I look forward to many more collaborative research projects with her in the years to come.

I would like to acknowledge the assistance of Dr. Akio Makashima, UCLA School of Engineering; John deGrosse, Ram Alkaly and Gerhardt Stummer, UCLA Department of Geology; the staff of Dr. Ian Kaplan's Group for use of the hydrothermal equipment, UCLA Department of Geochemistry, Tom Kaufman for the first hydration experiment that worked, Karen Robinson-Bild for developing SPECTRA 3 for generating punched output, and Jeannie Sells and Nila, UCLA Institute of Geophysics, for drafting and typing of manuscripts.

For the submission of samples and personal communications on their archaeological association, I would like to thank Drs. David Herrod and Frank Norrick for giving us a great working environment for our month's stay at Lowie Museum of Anthropology, University of California, Berkeley; the late Professor Robert Heizer, Department of Anthropology, University of California, Berkeley, for his permission to remove samples from obsidian artifacts and to radiocarbon date numerous organic materials; Dr. Dave

Frederickson, Chairman, Department of Anthropology, California State College at Sonoma, and Vera-Mae Frederickson for their hospitality during our stay in Berkeley; and Dr. James A. Bennyhoff of the same institution who introduced me to the problems and complexity of California prehistory; J. W. Michels, University of Pennsylvania; Dr. William Clewlow, Chris White, and Leonard Nelson, UCLA Department of Anthropology; Ward Upson, Santa Rosa, and Robert L. Orlins, Department of Anthropology, University of California, Davis.

I would like to thank Academic Press for permission to reproduce a chapter from Exchange Systems in Prehistory.

The author is most grateful to Professor Frank Findlow, Dept of Anthropology, Columbia University for reviewing portions of the manuscript and assistance in programming.

Writing acknowledgements leaves one searching days past and the corners of the brain to find all those who have helped; to you I say: Thank you, it was great.

TABLE OF CONTENTS

	Page
Preface	ii
Acknowledgements	iii
LIST OF TABLES	x
LIST OF PLATES	xii

PART 1: OBSIDIAN HYDRATION DATING

Chapter

1   EMPIRICAL OBSIDIAN HYDRATION RATES FOR CALIFORNIA 2

      INTRODUCTION 2

      THE SAMPLING OF OBSIDIAN SOURCES 4

      CHEMICAL CHARACTERIZATION OF SOURCES 5

         Introduction 5
         Short Half-Life Neutron Activation Analysis 8
         Long Half-Life Radionuclide Neutron Activation Analysis 9

      DATA COLLECTION AND MEASUREMENT FOR HYDRATION RATES 19

         Selection Strategies 19
         Selection Criteria 20
         Measurements for Hydration Rates 21

      EMPIRICAL OBSIDIAN HYDRATION RATES FOR CALIFORNIA 24

         Source-Specific Hydration Rates 24
         Inforporation of Temperature as a Variable in Empirical Hydration Rates 28
         An Evaluation of Physical Models and Empirical Equations Describing the Hydration Process 31

      CONCLUSIONS 37

Chapter		Page
2	HYDROTHERMALLY-INDUCED HYDRATION EXPERIMENTS	41
	INTRODUCTION	41
	THE EXPERIMENTS	41
	CALCULATION OF INDUCED HYDRATION RATES	48
	CONCLUSIONS	49
3	INTRINSIC VARIABLES OF THE HYDRATION PROCESS	52
	INTRODUCTION	52
	EMPIRICAL INTRINSIC VARIABLES	57
	Variables 1-8: Weight Percent of the Eight Major Oxides	57
	Variable 9: Density	58
	Variable 10: Hardness	59
	Variable 11: Maximum Water Concentration in the Hydration Layer	63
	Variable 12: Internal Water Concentration	66
	Variables 13-14: The Structural Bonding of Water	67
	Variable 15: Degree of Crystallinity	73
	CALCULATED INTRINSIC VARIABLES	73
	Variables 16-18: Silicon-Oxygen Ratio, Oxygen Activity, and Structural Factor	74
	Variable 19: Alumina Factor	80
	Variable 20: Specific Volume	80
	DETERMINATION OF THE INTRINSIC VARIABLES OF THE HYDRATION PROCESS	82
	DISCUSSION	86

PART 2: REGIONAL EXCHANGE SYSTEMS

4	INTRODUCTION TO THE STUDY OF REGIONAL EXCHANGE SYSTEMS	93
	INTRODUCTION	93
	THE SPACE-UTILITY FUNCTION OF EXCHANGE	95
	THE CONTEXT OF STUDYING EXCHANGE IN CALIFORNIA	99
5	THE REGIONAL EXCHANGE OF OBSIDIAN IN THE LATE HORIZON IN CALIFORNIA	104

Chapter		Page
	INTRODUCTION	104
	A DESCRIPTION OF EXCHANGE SYSTEMS	106
	A QUALITATIVE ANALYSIS OF CERTAIN SYSTEMIC VARIABLES	110
	CONCLUSIONS	114
6	SOURCE-SPECIFIC EXCHANGE SYSTEMS: A COMPARATIVE SYSTEMS ANALYSIS	117
	INTRODUCTION	117
	METHODOLOGICAL FRAMEWORK	117
	NOTES AND DESCRIPTION OF EXCHANGE SYSTEMS	118
	Obsidian Butte Exchange System	119
	Bodie Hills Exchange System	121
	Casa Diablo Exchange System	123
	Fish Springs Exchange System	126
	Annadel Exchange System	128
	St. Helena Exchange System	130
	Clear Lake Exchange System	133
	Medicine Lake Exchange System	135
	Surprise Valley Exchange System	138
	Coso Exchange System	140
	COMPARATIVE SYSTEMS ANALYSIS	143
	DISTANCE AS A VARIABLE	145
	QUANTITATIVE SYSTEMS ANALYSIS	149
	CONCLUSIONS	152
7	EXCHANGE SYSTEM GROWTH AND CHANGE	154
	INTRODUCTION	154
	MECHANISM OF FORMATION, STABILITY, AND CHANGE OF EXCHANGE SYSTEMS	154
	DIACHRONIC PRODUCTION RATES OF THREE INTERRELATED EXCHANGE SYSTEMS	159
	The St. Helena System Viewed from CCo-30, CCo-309, and CCo-308	159
	Diachronic Production Rate at Bodie Hills Obsidian Quarry	161
	Diachronic Production at the Casa Diablo Obsidian Quarry: A View from Mno-382	167
	Three Interrelated Exchange Systems: A Discussion	168

Chapter		Page
	EVOLUTION OF SYSTEMS AND THEIR RESPONSES ...	172
8	SUMMARY AND CONCLUSIONS	174
APPENDICES		177
1	FURTHER NOTES ON OBSIDIAN SOURCES............	178
2	PRIMARY DATA SET FOR SOURCE-SPECIFIC OBSIDIAN HYDRATION RATES	183
2A	CHEMICAL CHARACTERIZATION AND SAMPLE PROVENIENCE DATA	185
2B	ASSOCIATED DATE AND HYDRATION MEASUREMENT DATA	206
BIBLIOGRAPHY		224

LIST OF TABLES

Table		Page
1-1	Geographic Designation of Obsidian Sources	7
1-2	Short Half-Life Radionuclide Analysis of Obsidian Source Sample Groups by Neutron Activation	10
1-3	Characterization Matrix of <u>Northern</u> California Obsidian Sources, Indicating the Sample Overlap of Short Half-Life Data Through Stepwise Discriminate Analysis	12
1-4	Characterization Matrix of <u>Central</u> California Obsidian Sources, Indicating the Sample Overlap of Short Half-Life Data Through Stepwise Discriminate Analysis	13
1-5	Characterization Matrix of <u>Southern</u> California Obsidian Sources Indicating the Sample Overlap of Short Half-Life Data Through Stepwise Discriminate Analysis	14
1-6	Parameters of Neutron Activation Analysis	15
1-7	Order of Discriminating Isotopes Used in Chemical Characterization of California Obsidian Sources	20
1-8	Data for Source-Specific Obsidian Hydration Rate Curves Condensed from Appendix 3	25
1-9	Empirical Source-Specific Obsidian Hydration Dating Equations Using the Mathematical Model: Log T = Log a + b Log X (Equation 1-4)	28
1-10	Geographical Location, Mean Annual Air Temperature, and Natural Effective Temperature for 120 Weather Stations in California (modified after Felton 1965)	33
1-11	Rate Constants of Some Obsidian Hydration Models	39
1-12	Degree of Fit of the Hydration Rate Models Measured by Pearson's R	40
2-1	Experimental Conditions of Induced Hydration Experiments	42
2-2	Results of 150°C Induced Hydration Experiments	43
2-3	Results of 163°C Induced Hydration Experiments	45

Table		Page
2-4	Results of $172^{\circ}C$ Induced Hydration Experiments	46
2-5	Results of $200^{\circ}C$ Induced Hydration Experiments	47
2-6	Coefficients for Source-Specific Induced Hydration Equations	50
3-1	Major Oxide Analysis of Obsidian Samples (Weight-Percent)	53
3-2	X-Ray Fluorescence Spectrographic Conditions	56
3-3	Physical Properties of Obsidian Samples	60
3-4	Petrographic Study of Crystallites in Obsidian Source Samples	68
3-5	Calculated Intrinsic Variables	75
3-6	Sample Calculation of Silicon-to-Oxygen Ratio, Factor S (after Huggins 1944)	81
3-7	Sample Calculation of Normalized Analysis (1-1-1-1) Activity Factor, O_A, and the Number of Broken Structural Bonds, R (after Carron 1969:table 2)	82
3-8	Sample Calculation of Factor A, Alumina Factor	83
3-9	Sample Calculation of Factor Ω, Specific Volume Factor	84
3-10	Sample Calculation of Na/Na+K	85
3-11	Summary Table of Chemical and Physical Properties and Calculated Factors for Obsidian Samples	87
3-12	Variables of the Hydration Process	91
4-1	Number of Groups Reporting Specific Exchange Items Mentioned in the Ethnographic Literature as Being Imported and Exported (after Davis 1961)	103
5-1	Reference Index to Data Used in SYMAP	108
6-1	Population Estimates of Ethnolinguistic Groups	146
6-2	Results of Multilinear Regression Analysis of Three Systemic Variables of the Ten Exchange Systems	151
6-3	The Characteristics of the Ten Egalitarian Exchange Systems in California at Q = 10%, Based on Stepwise Regression Analysis	152

LIST OF PLATES

Plate		Page
1-1	Map of Obsidian Sources.	6
1-2	Characterization Matrix	18
1-3	Natural Effective Temperature Map of California.	30
1-4	Diagram of Natural Effective Temperature as a Function of Depth in Soil.	32
3-1	Water Profiles of Several Hydration Bands	64
4-1	Systemic Work Expenditure in Resource Acquisition by Direct Access and Linear Exchange.	98
5-1	A Synagraphic Map of the Prehistoric Egalitarian Exchange Systems in the Late Horizon of California	107
5-2	A Demonstration of the Correspondence of the Major Trails and Gradients of the Distributions of Obsidian within the Exchange Systems	112
5-3	A Demonstration of the Effects of Alternative Regional Resources on the Distribution of Obsidian within the Exchange Systems	113
5-4	A Demonstration of Lack of Correspondence Between the Boundaries of Ethnolinguistic Groups and Changes in the Distributions of Obsidian within the Exchange Systems	115
6-1	Obsidian Buttes Exchange	120
6-2	Bodie Hills Exchange	122
6-3	Casa Diablo Exchange	124
6-4	Fish Springs Exchange	127
6-5	Annadel Exchange	129
6-6	St. Helena Exchange	131
6-7	Clear Lake Exchange	134
6-8	Medicine Lake Exchange.	136
6-9	Surprise Valley Exchange	139
6-10	Coso Exchange	141

Plate		Page
6-11	Spatial Separation of the Bodie Hills and Casa Diablo Exchange Systems	144
6-12	Differential Utilization of Lithic Materials for Making Stone Tools within the "Supply Zones" and the "Contact Zones," as defined by Renfrew, Dixon, and Cann 1968	148
6-13	Measurement of the Effective Population and Distance from a Source to a Point of Consumption	150
7-1	A Hypothetical Case of How the Increased Evaluation of a Material Can Support Its Consumption within an Exchange System	157
7-2	St. Helena Exchange System Obsidian Production Rate Curve	162
7-3	Synagraphic Map of the Debitage at the Bodie Hills Obsidian Source, California	163
7-4	Bodie Hills Exchange System Obsidian Production Rate Curve	165
7-5	Bodie Hills Lithic Production Modes	166
7-6	Casa Diablo Exchange System Obsidian Production Rate Curve	169
7-7	Diachronic Production Rates of the Three Exchange Systems under Study	170

PART 1

OBSIDIAN HYDRATION DATING

There is much to be learned about matter, particularly crystal formation, while glasses are anomalous within our usual conceptualization of what we call "the states of matter".

Chapter 1

EMPIRICAL OBSIDIAN HYDRATION RATES FOR CALIFORNIA

INTRODUCTION

The obsidian hydration dating technique has been shown to be useful to archaeology (Katsui and Kondo 1965; Michels 1967, 1973; Meighan et al. 1968; Johnson 1969; Suzuki 1973; Bell 1977; Singer and Ericson 1977) and geology (Friedman 1968; Friedman and Peterson 1971; Friedman et al. 1973). Currently, there can be problems in the accuracy of the technique due to uncontrolled variability of several important variables of the hydration process. The hydration phenomenon involves the development of a measurable birefringence stress layer through a sequence of processes which are not totally understood. Atmospheric water is chemically absorbed on the surface of obsidian. This water diffuses into the interior of the obsidian as functions of time and temperature (Friedman and Smith 1960). It is also feasible on theoretical grounds (Ericson 1973a; Ericson, MacKenzie, and Berger 1976) that the water also reacts with the structure which causes the hydration rate to deviate from the diffusion model proposed by Friedman and Smith (1960) (Ericson 1975). This deviation, observed as a retardation of the rate as function of time, may have been observed for a proposed obsidian hydration rate for the American Southwest (Findlow et al. 1975).

In the absence of a complete understanding of the hydration process and the variables controlling the rate of hydration, there has been considerable debate over the mathematical form based on archaeological evidence (Clark 1961a, 1961b, 1964; Meighan, Foot, and Aiello 1968; Meighan 1970; Johnson 1969; Friedman and Smith 1960), the physical mechanism of the hydration process (Marshall 1961; Haller 1963; Friedman, Smith, and Long 1966; Ericson 1975; Ericson, MacKenzie, and Berger 1976), and the variables which influence the hydration rates (Friedman and Smith 1960; Aiello 1969; Ericson 1973a, 1975; Kimberlin 1971; Ericson and Berger 1976; Kimberlin 1976; Friedman and Long 1976; Ambrose 1976). As originally formulated by Friedman and Smith (1960) the obsidian hydration dating technique relied on a general diffusion equation having two variables, namely, time and temperature. To facilitate the application of the technique, broad temperature zones were established after the work of Chang (1957), within which a zonal hydration rate was to be used. Later, based on archaeological evidence, Clark (1961a, 1961b, 1964) and Meighan, Foote, and Aiello (1968) suggested that the proposed diffusion model did not fit the empirical hydration data. In support of their original thesis, Friedman, Smith, and Long (1966) defended their diffused model of hydration with the results of a four-year induced hydration experiment. The impact of these findings was to suggest to researchers that tighter data control was definitely necessary in order to resolve the hydration problem. A summary discussion of subsequent regional studies has been presented in Ericson, MacKenzie, and Berger 1976:39.

Even with increased geographical control, yet another form of variability was observed in hydration rate formation. Although Friedman and Smith (1960) did demonstrate hydration rate differences between trachytic and rhyolitic obsidian families, they did not suggest the degree of importance of chemical factors within each family of obsidians. As a result, a series of papers now suggests the importance of chemical composition in affecting hydration rates (Aiello 1969; Ericson 1969, 1973a; Ericson and Berger 1976; Kimberlin 1971, 1976; Michels and Bebrich 1971; Morgenstein and Riley 1973, 1975; Suzuki 1973; Layton 1973; Ambrose 1976; Friedman and Long 1976).

In summary, prior research has continued to refine the obsidian hydration dating technique by determining and controlling variables of the hydration process which have been defined as time, temperature, and chemistry of the obsidian. We are not yet assured of the actual mathematical model which best fits the hydration process: the Friedman School suggests and supports Fick's Second Law of Diffusion and the Meighan School suggests an empiricist's approach to modeling the archaeological and hydration data. Fortunately, in a broader perspective of research and development, this debate has stimulated the continuation of research efforts so necessary in resolving complex problems.

As part of the above research effort, the first part of this monograph, part 1, is designed to promote a further understanding of the chemical and physical properties of obsidian, preliminarily defined by Ericson et al. (1975) These properties, in turn, will be examined to define their influence on the rates of hydration which are determined from both empirical (archaeological) data and experimental (laboratory) data. Accordingly, part 1 is divided into three chapters. The first chapter describes the procedures and results of defining source-specific empirical obsidian hydration rates, based on well-controlled archaeological data. The second chapter explores the experimentally induced hydration process under controlled laboratory conditions. These results are compared with those from archaeological data to determine the degree of fit between data sets. The third chapter describes the procedure and results of measuring a wide suite of chemical and physical properties, as well as the logic involved in generating a set of calculated factors. Finally, the changes in the hydration rates are examined in terms of changes among the properties of the respective obsidian sources. This multi-level approach to understanding the medium, obsidian itself, will provide a basic understanding of the nature of the hydration process. Part 1 attempts to answer some of the following basic questions:

1. What chemical and physical properties characterize rhyolitic obsidian?

2. What is the degree of variation of hydration rates among rhyolitic obsidian sources?

3. What mathematic model, if any, best fits the hyration data?

4. Does the laboratory-induced process reduplicate the "natural hydration" process?

5. What intrinsic properties of rhyolitic obsidian are important in affecting the observed changes in hydration rates?

6. What do these observed factors indicate about the hydration process?

The above questions are discussed in the three chapters that follow.

Chapter 1 describes the procedures and results of defining source-specific empirical obsidian hydration rates. The first step in defining these rates is to locate and sample known obsidian sources in California, Western Nevada, and Southern Oregon. The details of the obsidian source program were recently published (Ericson, Hagan, and Chesterman 1976). Secondly, these source samples are chemically analyzed by extensive instrumental neutron activation analyses and some X-ray fluorescence analyses.

In turn, the distinguishing characteristics of these data are defined through multivariate statistical analysis, resulting in the chemical characterization of each obsidian source. These two steps provide a means to accurately identify the original source of an obsidian artifact. A large sample of obsidian artifacts and associated radiocarbon dates were obtained through the courtesy of many individuals and several museums. Both types of samples were analyzed by applying chemical characterization and radiocarbon dating analysis which form source-specific data sets.

These data sets are analyzed by linear regression to give the coefficients of a generalized mathematical equation describing the best fit to the data. A set of empirical equations, which have been previously used to describe the hydration process, are examined to suggest a general mathematical model of the hydration process. The objective in chapter 1 is to describe a series of source-specific empirical hydration rates for a number of sources used by aboriginal people. These equations will be useful for determining obsidian hydration dates in California. The details of the procedure and the results are presented in the pages that follow.

THE SAMPLING OF OBSIDIAN SOURCES

Beginning in 1970, it was recognized that available obsidian source collections were insufficient for the intended research program. The primary objective of the field studies which were conducted was to gain samples from each obsidian source for measurement of their chemical and physical properties. This aspect of research was continued from September 1970 until August 1975.

The strategy employed in establishing the collections was to physically locate a given obsidian source through information provided in the literature or in the field by examining the alluvium and drainage of a given volcanic field. Once located in the field, the structure of the obsidian source was sampled at multiple, designated locations, in order to eventually determine chemical homogeneity of the structure. The collections and field notes are accessioned both in the Obsidian Hydration Dating Laboratory, Department of Anthropology, and the Museum of Geology, Department of Geology, UCLA.

In the course of geological surveying, small collections of debitage within prehistoric quarry workshops were generally made with the intention of comparing inter-source lithic technology and production. However, the inadequacy of the sampling procedures and brevity field observations foreclose any idea of making comparisons. These collections are accessioned at the Museum of Anthropology, UCLA.

The report for the period 1970-73, on the initial phase of research on the obsidian sources in California, Western Nevada, and Southern Oregon, has been completed and appears elsewhere (Ericson, Hagan, and Chesterman 1976). Further notes on the obsidian sources, which update the information contained in the original report (ibid.), are presented as an appended section (appendix 1).

The obsidian sources which have been studied are shown in plate 1-1, and their geographical locations are presented in table 1-1. The samples from these sources constitute the materials which have been extensively examined by chemical and physical measurement. The details of this work are discussed later. In all, there are more than two dozen obsidian sources in the region of study (cf. appendix 1). Several potential sources remain to be reported upon, and some reported in the literature have been discounted as non-obsidian sources (Ericson, Hagan, and Chesterman 1976; appendix 1).

With the sampling completed, the immediate problem was to establish chemical criteria by which the source of an unknown obsidian artifact could be identified. This technical capability would allow the obsidian hydration data or other data, e.g., exchange data, to be categorized into source-specific groups. With the availability of the technology, support, and facilities at UCLA, the obsidian source samples were chemically characterized.

CHEMICAL CHARACTERIZATION OF SOURCES

Introduction

Chemical characterization can be defined as a process or procedure which defines the chemical parameters by which a set of sources can be distinguished. The function of chemical characterization as applied to archaeology has been to accurately identify the origin or source of artifacts. There is an extensive literature on this subject which has been reviewed by Perlman, Asaro, and Michel (1972) with regards to selective techniques of chemical analysis, techniques of grouping chemical data, and types of materials which have been characterized.

The analysis and chemical characterization of obsidian sources in California are not without precedent (Griffin, Gordus, and Wright 1969; Jack and Carmichael 1969; Stevenson, Stross, and Heizer 1971; Bowman, Asaro, and Perlman 1973; Jackson 1974; Jack 1976). Although this information serves as a foundation, the present work attempts to complete the original problem of accurate and comprehensive obsidian source characterization in California proposed by Jack and Carmichael (1969).

In this particular chemical characterization study, which was necessary for developing source-specific hydration rate curves, three procedures of

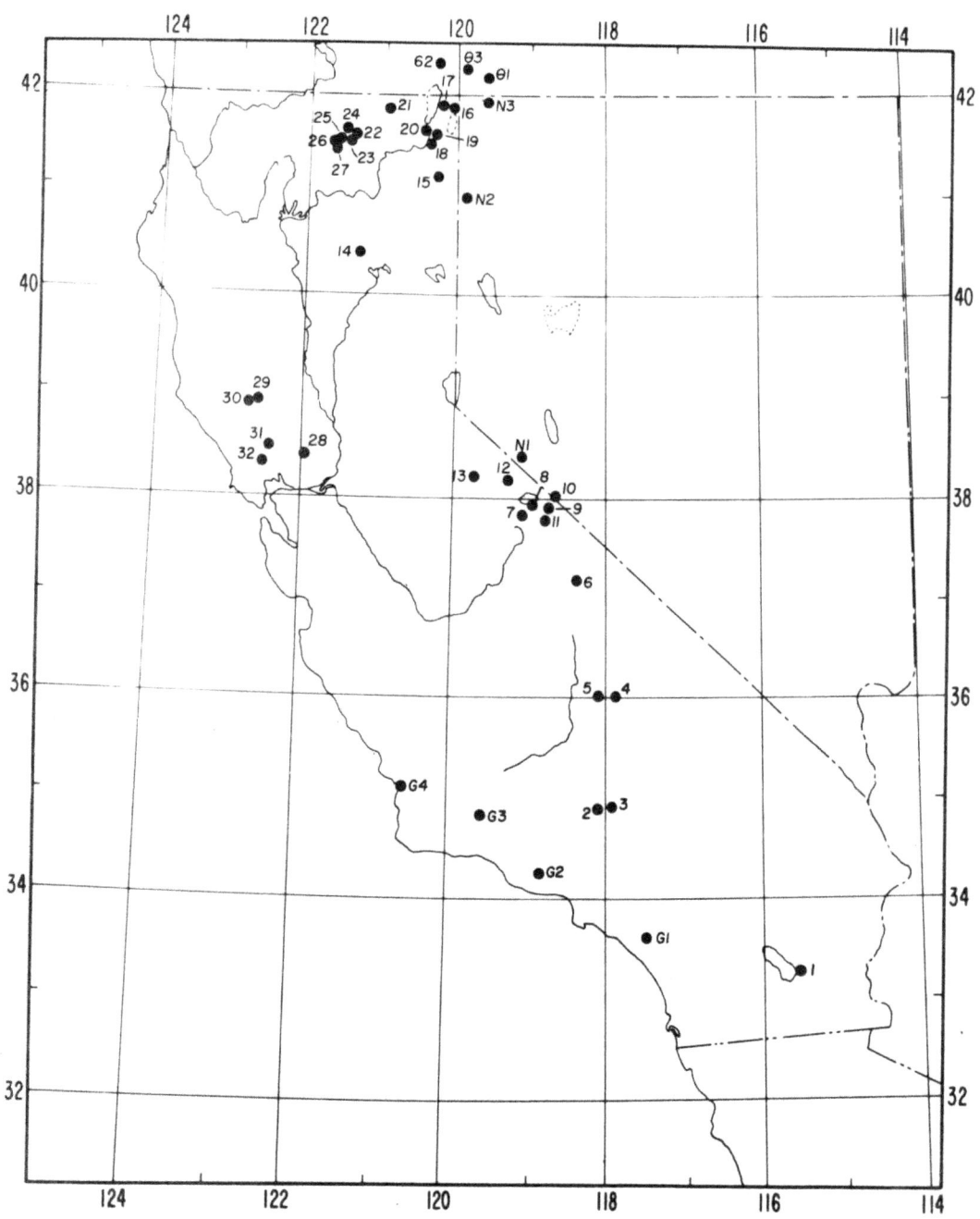

Plate 1-1. Map of Obsidian Sources. (Refer to table 1-1 for name of each source.)

Table 1-1

GEOGRAPHIC DESIGNATION OF OBSIDIAN SOURCES
(cf. plate 1-1)

California Obsidian Sources

1. Obsidian Butte, Imperial County
2. Emerald Mountain, Kern County
3. Jawbone Canyon, Kern County
4. Sugarloaf, Inyo County
5. Monache Meadows, Tulare County
6. Fish Springs, Inyo County
7. Inyo Craters, Mono County
8. Mono Craters, Mono County
9. Mono Glass Mountain, Mono County
10. Truman Canyon-West Queen Mine, Mono County
11. Casa Diablo, Mono County
12. Bodie Hills, Mono County
13. Levitt Peak, Tuolumne County
14. Deer Creek, Tehama County
15. Jess Valley, Modoc County
16. Cowhead Lake, Modoc County
17. 8-Mile Creek, Modoc County
18. Buck Mountain, Modoc County
19. Fandango Valley, Modoc County
20. Sugarhill, Modoc County
21. Steele Swamp, Modoc County
22. Dacite-Rhyolite Composite Flow, Glass Mountain, Modoc County
23. Rhyolite Obsidian Flow, Glass Mountain, Siskiyou County
24. Cougar Butte, Siskiyou County
25. Medicine Lake Glass Flow, Siskiyou County
26. Little Glass Mountain, Siskiyou County
27. Grasshopper Flat, Siskiyou County
28. Winters, Solano County
29. Borax Lake, Lake County
30. Mount Konocti, Lake County
31. Napa Glass Mountain, Napa County
32. Annadel Farms, Sonoma County

California Sources
(Natural Glasses)

G1. El Toro Glass, Orange County
G2. Grimes Canyon Fuse Shale, Ventura County
G3. Cuyama Glass, Santa Barbara County
G4. Shell Beach Zeolitized Tuffs, San Luis Obispo County

Western Nevada Obsidian Sources

N1. Mount Hicks, Mineral County
N2. Duck Flat, Washoe County
N3. Long Valley, Washoe County

Southern Oregon Obsidian Sources

O1. Beatty's Butte, Lake County
O2. Glass Mountain, Lake County
O3. Glass Butte, Lake County

chemical analysis were used: X-ray fluorescence analysis and instrumental neutron activation analysis of both short and long half-life radionuclides.

At first, X-ray fluorescence analysis of the sources coupled with a ternary or tri-poled grouping of the elementary composition of zirconium, rubidium, and strontium was tried after Jack and Carmichael (1969) since it promised to be non-destructive, the least toxic, most rapid, and least expensive means of approaching the problem. However, with the problem of chemical overlap in the numbers of sources involved, which increased the likelihood of statistical β-error, this original procedure was soon abandoned in favor of instrumental neutron activation analysis of short half-life radionuclides. It is important to say that the original X-ray fluorescence technique and recent modifications (Jackson 1974; Jack 1976) are useful but their powers of discrimination impose definite limitations. Even attempts at characterization using short half-life radionuclides through instrumental neutron activation analysis (INAA) did not always overcome the inherent problem of chemical overlap. In fact, this procedure, which is described more fully in the following pages, requires the geographical grouping of chemical data. It appears that INAA of long half-life radionuclides has the best promise of overcoming the overlap problem. With the adoption of multivariate statistical analysis in the form of stepwise discriminate analysis and long half-life radionuclide INAA, a new and powerful methodology for chemical characterization has been developed and tested.

Short Half-Life Neutron Activation Analysis

Short half-life radionuclide instrumental neutron activation analyses were performed to analyze the trace chemistry of the obsidian source samples, using the following procedure. Each obsidian sample was prepared by cutting a small wedge sample with a Felke Di-Met saw with a $3\frac{1}{2}$-inch diamond-charged brass saw blade, weighing each 5-20 mg sample to 10^{-5} gram accuracy with a Mettler analytical balance, transferring it into a $\frac{1}{2}$-dram, labeled, snap-top polyethylene tube for irradiation. Seven samples and two Napa Glass Mt. powder samples, in-house neutron flux standards, were packed into a larger polyethylene tube for irradiation. The samples were irradiated, using the "rabbit" system, at a flux of 2×10^{12} neutrons/cm^2/sec in the reactor of the Nuclear Engineering Facility, UCLA. The irradiation time for the short half-life elements was three to five minutes. The irradiated samples were then rushed to the analyzing unit located one-quarter mile away. Prior to a given experiment, a set of gamma-ray standards were used to standardize the analyzing unit which consisted of a lithium-drifted germanium detector whose efficiency is 6%, relative to a 3" x 3" NaI (Tl) detector at a source-to-detector distance of 25 cm and with a resolution of 4.5 Kev for the 1332 Kev peak of Co^{60}. Gamma-ray data was discriminated by a Nuclear Data 2200 4096 channel analyzer, which, in turn, was read out on a magnetic tape assembly for subsequent computer analysis. The computer program, SPECTRA, developed by Baedecker (1976), was used to analyze the gamma-ray spectrum data of the following short half-life elements of each sample: dysprosium (Dy), manganese (Mn), sodium (Na), and potassium (K).

Standardized ratios of the values of these elements of the unknown sample and the Napa Glass Mt. neutron flux standards were received as output of the SPECTRA program. In turn, these standardized ratios were converted by multiplication to quantitative elemental values using the following factors: Dy, 2.970 ppm; Mn, 169.9 ppm; Na, 3.609 wt %, K, 3.929 wt %. The factors represent the mean values of the Napa Glass Mt. standard relative to the published analysis of USGS G-2 standard (Flanagan 1973).

The short half-life neutron activation analysis of samples was conducted irregularly over a two-year period beginning in November 1971. Generally, these analyses were performed in conjunction with Jerome Kimberlin, Departments of Chemistry and Anthropology, UCLA. The irradiations of the samples were done under the supervision and license of Professor John Wasson, Department of Chemistry, UCLA. By conducting experiments concordant with the irradiation schedule of Professor Wasson's research group, the cost of sample irradiation was minimized.

The results of the short half-life neutron activation analysis are presented in table 1-2. Table 1-2 presents the mean and standard deviation of each analyzed element and sample size for each obsidian source group.

At this point in the analysis, the complexity of the data set, presented in table 1-2, necessitated a unique solution to the problem of statistical differentiation of the chemical data. For the first time, in 1973, at the suggestion of Professor Dwight Read, stepwise discriminate analysis was employed to discriminate inter-source chemical variability. The application of this multivariate statistical analysis provided an _a priori_ technique to evaluate chemical variability, usually determined by simple comparison (Gordus, Wright, and Griffin 1968) or simple plotting (Jack and Carmichael 1969).

As a trial, all the data presented in table 1-2 was entered into the stepwise discriminate program, BMDΦ7M with F-to enter and F-to-remove set at 4.0 and 3.0 respectively. It was immediately apparent that there was significant chemical overlap among some of the sources. This was rather a disconcerting discovery, ipso facto, since Gordus, Wright, and Griffin (1969) had already shown the power of discrimination by two variables alone, namely, Na/Na and Mn. As a resolution to this inherent problem, the source data was grouped by region and reanalyzed by stepwise discriminate analysis. Again, the results of stratifying the data in this fashion indicate some chemical overlap among Northern California (table 1-3), Central California (table 1-4), and Southern California (table 1-5) obsidian sources.

It is now possible on a routine basis to chemically characterize obsidian artifacts if their regional provenience is known. Notwithstanding the problem of chemical overlap, stepwise discriminate analysis does indicate the nature of the statistical overlap by the utilization of posterior probabilities and Mahalabinos D^2 statistic. As a result of this work, three "packaged" programs for artifact characterization are now available from the Obsidian Hydration Dating Laboratory, Department of Anthropology, UCLA.

Long Half-Life Radionuclide Neutron Activation Analysis

Long half-life radionuclide instrumental neutron activation analyses were performed to analyze the trace chemistry of the obsidian source samples

Table 1-2

SHORT HALF-LIFE RADIONUCLIDE ANALYSIS OF OBSIDIAN SOURCE SAMPLE GROUPS BY NEUTRON ACTIVATION

Obsidian Source	Number	Dy (ppm)	μn (ppm)	Na Wt. %	K Wt. %
Annadel	7	3.09 ± .38	356 ± 16	3.898 ± .52	3.228 ± .506
Benton	9	1.31 ± .18	719 ± 52	3.231 ± .234	4.054 ± .467
Bodie Hills	6	n.d.	443 ± 10	2.974 ± .087	3.670 ± .312
Borax Lake	8	2.81 ± .30	173 ± 12	2.908 ± .212	4.013 ± .499
Buck Mt.	25	1.05 ± .13	419 ± 17	3.086 ± .125	4.347 ± .466
Casa Diablo	4	1.02 ± .22	302 ± 11	2.854 ± .033	4.324 ± .395
Coso	5	3.44 ± .10	257 ± 4	3.336 ± .156	4.018 ± .253
Cowhead Lake	19	1.94 ± .25	928 ± 62	3.683 ± .210	3.734 ± .491
Fish Springs	8	1.83 ± .39	868 ± 37	3.412 ± .125	4.123 ± .793
Grimes Canyon	0	---	---	---	---
Inyo Craters	17	1.78 ± .21	511 ± 39	3.572 ± .290	4.423 ± .520
Jawbone Canyon	5	n.d.	315 ± 15	2.955 ± .345	2.709 ± 1.933
Modoc Glass Mt.	11	1.56 ± .29	315 ± 32	3.366 ± .266	4.174 ± .355
Monache Meadows	4	n.d.	501 ± 6	3.668 ± .708	4.145 ± 1.053
Mono Craters	9	1.84 ± .27	375 ± 30	3.209 ± .265	4.336 ± .487
Mono Glass Mt.	8	2.42 ± .66	304 ± 13	3.069 ± .178	4.168 ± .290
Mt. Konocti	10	2.26 ± .25	234 ± 29	2.872 ± .190	4.171 ± .471
Napa Glass Mt.	24	2.95 ± .29	168 ± 9	3.505 ± .214	3.994 ± .273
Obsidian Butte	8	7.88 ± .60	411 ± 17	3.797 ± .145	4.139 ± .391
Rustler Canyon	4	2.95 ± .02	772 ± 28	3.777 ± 108	3.169 ± 1.481
Shell Beach	0	---	---	---	---
Shoshone	6	n.d.	313 ± 17	2.625 ± .056	2.731 ± 1.990
Source X	0	---	---	---	---
Sugar Hill	21	1.21 ± .18	364 ± 22	2.952 ± .151	---

(contd.)

Table 1-2 (continued)

Obsidian Source	Number	Dy (ppm)	Mn (ppm)	Na Wt. %	K Wt. %
Duck Flat, Nev.	4	3.65 ± 1.72	395 ± 92	3.538 ± .262	4.414 ± .427
Fletcher, Nev.	4	1.86 ± 1.32	361 ± 91	2.913 ± .318	3.016 ± 1.353
Mt. Hicks, Nev.	4	2.20 ± 1.05	391 ± 6	3.053 ± .094	3.930 ± .239
Pine Grove Hills, Ore.	4	n.d.	502 ± 51	3.118 ± .235	2.966 ± 1.578
Beatty's Butte, Ore.	3	1.07 ± .01	442 ± 20	2.967 ± .072	4.841 ± .521
Long Valley, Nev.	3	3.69 ± .37	483 ± 15	3.648 ± .104	4.151 ± .464
Glass Mt., Ore.	2	3.97 ± .28	796 ± 5	4.076 ± .015	4.270 ± 1.696
Glass Butte, Ore.	3	2.85 ± .21	345 ± 4	3.379 ± .053	3.459 ± .104

Table 1-3

CHARACTERIZATION MATRIX OF NORTHERN CALIFORNIA OBSIDIAN SOURCES,
INDICATING THE SAMPLE OVERLAP OF SHORT HALF-LIFE DATA
THROUGH STEPWISE DISCRIMINATE ANALYSIS

		A	B	C	D	E	F	G	H	I
Buck Mt.	(A)	8								
Sugarhill	(B)		10							
Long Valley, Nev.	(C)	2		22						
Cowhead Lake	(D)				7					
Modoc Glass Mt.	(E)					6				
Duck Flat, Nev.	(F)	1					3			
Beatty's Butte, Ore.	(G)							3		
Glass Mt., Ore.	(H)								2	
Glass Butte, Ore.	(I)									3

Table 1-4

CHARACTERIZATION MATRIX OF CENTRAL CALIFORNIA OBSIDIAN SOURCES, INDICATING THE SAMPLE OVERLAP OF SHORT HALF-LIFE DATA THROUGH STEPWISE DISCRIMINATE ANALYSIS

		A	B	C	D	E	F	G	H	I	J	K	L	M	N
Borax Lake	(A)	8													
Mt. Konocti	(B)		9												
Napa Glass Mt.	(C)	3		21											
Annadel	(D)				7										
Bodie Hills	(E)					6									
Casa Diablo	(F)						4								
Mt. Hicks	(G)							3							
Benton Valley	(H)								9						
Pine Grove Hills, Ore.	(I)					1	1	1		2					
Mono Glass Mt.	(J)										7				
Mono Craters	(K)											8			
Inyo Craters	(L)											1	16		
Fish Springs	(M)													8	
Fletcher, Nev.	(N)					1	1								2

Table 1-5

CHARACTERIZATION MATRIX OF SOUTHERN CALIFORNIA OBSIDIAN SOURCES, INDICATING THE SAMPLE OVERLAP OF SHORT HALF-LIFE DATA THROUGH STEPWISE DISCRIMINATE ANALYSIS

		A	B	C	D	E	F	G	H	I	J	K	L
Fish Springs	(A)	8											
Coso	(B)		5										
Monache Meadows	(C)			4									
Jawbone Canyon	(D)				2								
Rustler Canyon	(E)					4							
Obsidian Butte	(F)						8						
Casa Diablo	(G)							4	2				
Mono Glass Mt.	(H)							1	7				1
Benton Valley	(I)									9			
Bodie Hills	(J)										6		
Mt. Hicks, Nev.	(K)											4	
Shoshone	(L)				1								2

14

Table 1-6

PARAMETERS OF NEUTRON ACTIVATION ANALYSIS

	Radionuclide (Element)	Photopeak Energy, Kev	Half-Life	Napa Flux Monitor Weight	USGS G-2, ppm.	Notes
(1)	Nd147	91.4	11.10D	38.9699 µg	Nd 60	None detected
(2)	Pa233	98.2	27.00D	18.23999 µg	Pa --	See (11)
(3)	Gd153	103.4	236.00D	4.01200 µg	Gd 5.0	Eu152 = ([4] + [9])/2
(4)	Eu152	121.7	12.20Y	0.48280 µg	Eu 1.5	Hf181 =([5] + [13])/2
(5)	Hf181	132.8	42.50D	6.70900 µg	Hf 7.35	
(6)	Ce141	145.2	33.00D	58.95999 µg	Ce 150	
(7)	Yb169	197.6	32.00D	3.72000 µg	Yb 0.88	
(8)	Ba131	216.0	12.00D	1234.89990 µg	Ba 1870	
(9)	Eu152	244.3	12.20Y	0.28280 µg	Eu --	See (4)
(10)	Tb160	298.5	72.10D	0.50420 µg	Tb 0.54	Th232=([2] + [11])/2
(11)	Th232	311.5	27.00D	18.23999 µg	Th 24.2	
(12)	Cr51	319.8	27.80D	24.04999 µg	Cr 7	
(13)	Hf181	481.8	42.50D	6.70900 µg	Hf --	See (5)
(14)	Cs134	604.2	2.05Y	15.25000 µg	Cs --	Problematic
(15)	Zr95	724.2	65.00D	239.39999 µg	Zr --	Problematic
(16)	Zr95	756.6	65.00D	239.39999 µg	Zr 300	
(17)	Nb95	765.6	35.00D	239.39999 µg	Nb --	
(18)	Cs134	795.6	2.05Y	15.25000 µg	Cs 1.4	Cs134 = ([14] + [18])/2
(19)	Sc46	889.4	83.90D	2.95100 µg	Sc 3.7	Sc46 = ([19]+ [23])/2
(20)	Rb86	1076.6	18.66D	187.09999 µg	Rb 168	
(21)	Fe59	1098.7	45.00D	0.92490 %	Fe 1.853	Fe59 = ([21] + [26])/2
(22)	Zn65	1115.4	245.00D	60.82999 µg	Zn 85	
(23)	Sc46	1120.0	83.90D	2.95100 µg	Sc --	See (19)
(24)	Co60	1172.6	5.26Y	10.69000 µg	Co 5.5	Co60 = ([24]+ [27])/2
(25)	Ta182	1220.8	115.00D	1.41500 µg	Ta 0.91	
(26)	Fe59	1290.9	45.00D	0.92490 %	Fe --	See (21)
(27)	Co60	1332.4	5.26Y	10.69000 µg	Co --	See (24)

in an attempt to resolve the problem of chemical overlap. Each obsidian sample was prepared by cutting a small rectangular sample with a Felke Di-Met saw with a $3\frac{1}{2}$ inch diamond-charged brass saw blade, labeling with diamond stylus, and weighing each 6-25 mg sample to 10^{-5} gram accuracy with a Mettler analytical balance. Ten samples, which constituted a group, were packed in ultra-pure aluminum soil for irradiation. The 20 sample group containers were packed together with six sealed Supersil quartz vials of powdered Napa Glass Mt. flux monitor standard, plus an iron wire flux monitor added to test linear flux variation during irradiation.

The Napa Glass Mt. flux monitor standard is the obsidian standard used in the laboratory for quantitative analysis. This standard is a ground, mixed, sieved, and standardized inside portion of an obsidian "bomb" from the western exposure of the Napa Glass Mt. obsidian source (Ericson, Hagan, and Chesterman 1976). The standardization was accomplished using the United States Geological Survey Granite Standard G-2, having the elemental values presented in table 1-6 (Flanagan 1973). The results of the standardization of the Napa Glass Mt. Standard are presented in table 1-6, courtesy of Jerome Kimberlin, Departments of Chemistry and Anthropology, UCLA (Kimberlin, personal communication).

The samples were irradiated an equivalent flux of 5×10^{13} neutrons/cm^2/sec for 23 hours at the University of Missouri nuclear reactor facility after which the irradiated samples were allowed to decay for three weeks for shipping back to UCLA under Atomic Energy Commission regulations.

A set of gamma-ray standards were used to standardize the analyzing unit which consisted of a Ge(Li), lithium-drifted germanium diode detector whose efficiency is 11%, relative to a 3 inch x 3 inch NaI (Tl) detector at a source-to-detector distance of 25 cm and with a resolution of 1.9 Kev for the 1332 Kev photo peak of Co^{60}. The detector was connected to a Nuclear Data 2200 multichannel (4096) analyzer equipped with a magnetic tape output. Analysis of the data was performed by using the computer program SPECTRA (Baedecker 1976) and modified by a version called ANSPEC with an option for punch-card output (Robinson-Bild, personal communication), using the neutron activation half-life and other parameters specified in table 1-6. The resultant program printout was checked by hand for errors and the punch-card output organized by source group for subsequent analysis. The quantitative analysis of the obsidian source samples was accomplished by comparison with the mean value of the Napa Glass Mt. Standard which has been standardized to the published analysis of USGS G-2 Standard (Flanagan 1973) as presented in table 1-6. These analyses were performed in conjunction with Jerome Kimberlin and with the assistance of Victoria Bennett and Terrence N. D'Altroy, UCLA. The irradiation and computing were done under the supervision and license of Professor John T. Wasson, UCLA, over a three-month period in 1975. The cost of transportation of the samples was defrayed by Professor Wasson on an exchange basis.

The results of quantitative analysis of the obsidian source samples, organized by source, is presented by Ericson and Kimberlin (1975). They present the name and sample size for all analyzed obsidian source groups and the group mean and standard deviation of each analyzed element. It is important to note that no effort was made to report the data either on a sample-by-sample or group-as-collected basis, simply because of the immense volume of the raw data, estimated to exceed 7 kg of computer paper. Thus, the potential of intra-source chemical characterization (cf. plate 1-2) remains as an exciting prospect which is beyond the scope of this present work. The " raw data" is available upon request and permission of the analysts.

With the completion and preliminary evaluation of the chemical analyses, the resolution of the basic problem was assured, i.e., whether or not these data could effectively discriminate among the obsidian sources. There was no reason to expect why, with the addition of 16 chemical variables as potential discriminators, these data would not eliminate the need of stratifying the chemical data into broad regional source groups. Again, stepwise discriminate analysis was selected to organize and analyze this enormous data set. However, first the ANSPEC punch-card data had to be edited for errors, arranged into source groups by hand sorting, double-checked, and coded with a sample group designation, which resulted in a 1380 card data deck- representing a 16 x 276 variable-to-sample matrix. In turn, the size of the matrix required a modification of the stepwise discriminate analysis program BMDP7M in order to accommodate the mass of data.

The result of stepwise discriminate analysis of these data is shown in plate 1-2. Plate 1-2 shows the chemical characterization matrix of obsidian source and sample points based upon long half-life radionuclide INAA data. It is apparent that little chemical overlap exists among the obsidian sources as indicated by the diagonal matrix of the source groups. The notable exceptions are the following: (1) the Fletcher source which appears to be a volcanic vent or intermixture of the Mt. Hicks and Bodie Hills sources, (2) Shell Beach Glass, which is not technically obsidian, and (3) although the obsidian sources show little or no chemical overlap, the as-collected sample groups, shown as the smaller intra-source group matrices in plate 1-2, do overlap. Perhaps the utilization of posterior probabilities and Mahalabinos D^2 statistics might distinguish unknown artifacts as intra-source members.

There are two additional points which are important to mention about the stepwise discriminate analysis. First, the order of the isotopes used in the process is described in table 1-7. The order and power of discrimination is important information to the analyst, since it specifies the elements which have the best success in actual discrimination. This order will vary as a function of the original data set, e.g., for California obsidian sources this order of elements is unique. Secondly, it is important to recognize the fact that stepwise discriminate analysis functions as a validating check on the original data set in the following way: the program forms original groups specified by the programmer, then, in turn, assigns individual samples to any available group. If an error had been made in the many different steps, it would have appeared as a misgrouped sample. Third and most importantly, stepwise discriminate analysis provides a means to objectively distinguish groups and statistically test the degree of overlap of rather large data sets. Finally, since the stratification of chemical groups has been tested, unknown artifacts can be classified and assigned membership to a particular group. It is at this point in a rather complex research project that the chemical characterization of artifacts can be routinely accomplished.

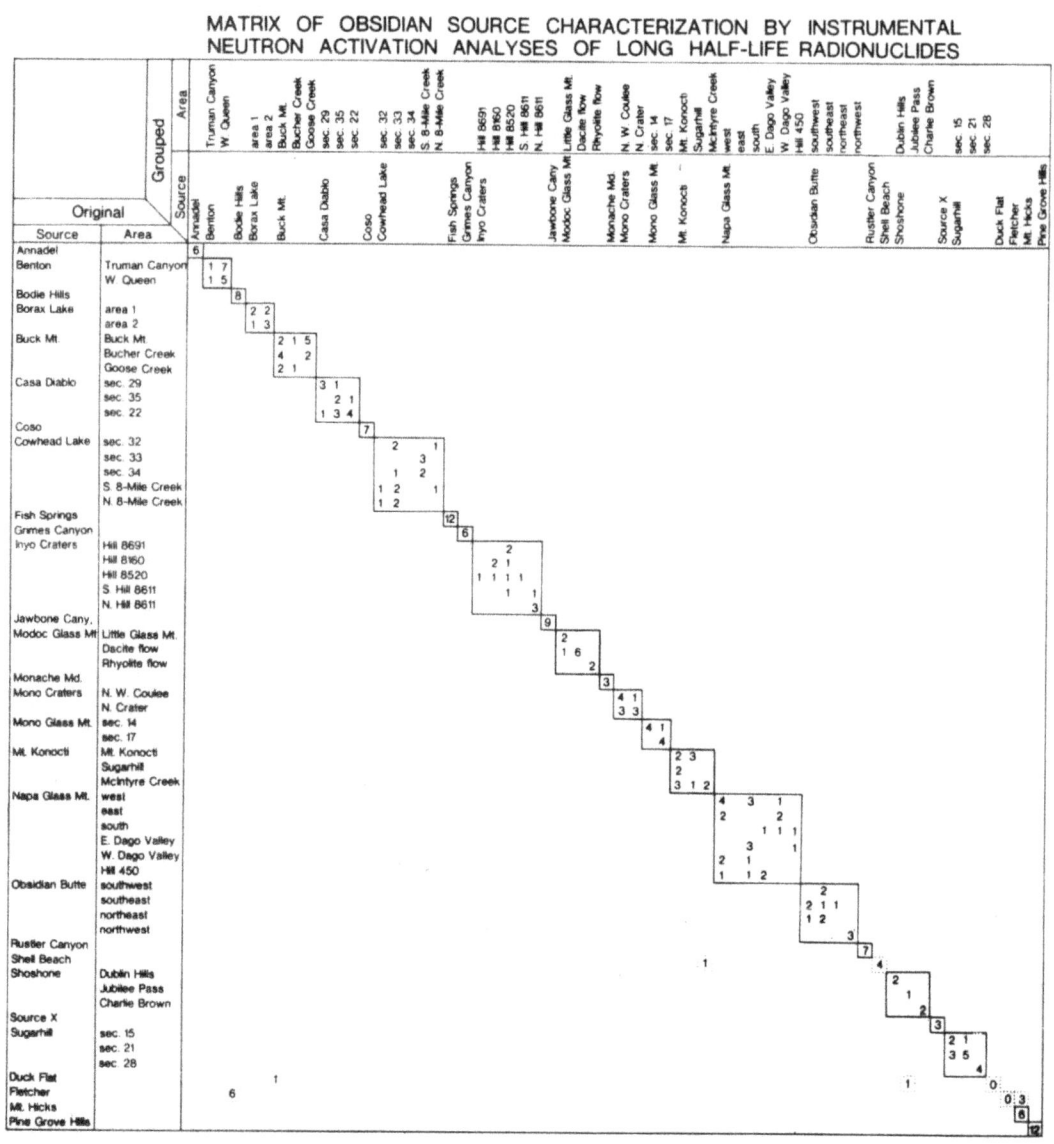

Plate 1-2. Characterization Matrix. These are the results of the stepwise discriminate analysis of the long half-life neutron activation analysis of the obsidian source samples. The numbers in the matrix denote the number of samples of a sampled source which were subsequently classified into a statistical group. A perfect discrimination would have resulted in a singularity matrix, i.e., all numbers would have been aligned on the diagonal. The solid boxes enclose a single geographical source of one or more sampling locations. The dotted boxes indicate that there is chemical and statistical overlap of source samples with other sources, which might result in the misclassification of unknown samples.

It is now possible on a routine basis to chemically characterize obsidian artifacts without knowing their regional provenience within California. This can be accomplished by using long half-life radionuclide instrumental neutron activation analysis of artifacts and the "packaged" program now available from the Obsidian Hydration Dating Laboratory, Department of Anthropology, UCLA. The methodology developed and tested here provides an objective means to chemically characterize obsidian artifacts by combining "long half-life" neutron activation and stepwise discriminate analysis.

In conclusion, the source or origin of an obsidian artifact can now be identified by applying three separate procedures depending upon the particular research strategy and the facilities available to the analyst. If the number of sources prehistorically utilized at a particular time or place is known to be small or chemically distinctive, then the analyst most likely will select the rapid-scan X-ray fluorescence technique (Jack and Carmichael 1969) or the short half-life radionuclide INAA technique proposed herein. However, if the requisite information is not always available to the analyst, the application of the long half-life radionuclide INAA technique of chemical characterization will provide the greatest precision and routinization of laboratory procedure.

DATA COLLECTION AND MEASUREMENT FOR HYDRATION RATES

With the completion of the sampling and chemical characterization of obsidian sources in California, source-specific obsidian hydration rates could now be developed by stratifying the temporally controlled obsidian hydration data by source. Early research (Aiello 1969; Findlow et al. 1975; Ericson 1975; Kimberlin 1976; Ericson and Berger 1976) had already demonstrated to a certain extent the refinement and increased accuracy involved in utilizing source-specific hydration rates. Moreover, a theoretical basis for such data control had been suggested (Ericson, MacKenzie, and Berger 1976). It was the manifestation of these empirical observations on the one hand and theoretical propositions on the other that provided the original incentive for conducting the entire research program.

Selection Strategies

Three separate strategies were employed to acquire samples for rate determinations. First, samples were solicited from ongoing excavations within Southern California over a five-year period. This method proved to be least effective in obtaining an adequate data set. The second strategy was to obtain from existing museum collections obsidian artifacts associated with existing radiocarbon dates using collated data (Ericson and Hagan n.d.). With the assistance of Professors Bennyhoff and Frederickson, this method was quite effective in obtaining a sizable number of controlled samples. The third and most effective means of obtaining samples was to search available museum collections by noting all organic samples, which would be useful as radiocarbon samples, and, in turn, all obsidian samples from a given collection. By matching the provenience of the organic and obsidian samples within the entire holdings of several museums, it was possible to assemble an adequate data set for hydration rate determination, e.g., two

Table 1-7

ORDER OF DISCRIMINATING ISOTOPES USED IN CHEMICAL
CHARACTERIZATION OF CALIFORNIA OBSIDIAN SOURCES

Step Number	Isotope
1	Ta182
2	Ce141
3	Yb169
4	Cs134
5	Th232
6	Eu152
7	Sc46
8	Fe59
9	Hf181
10	Co60
11	Tb160
12	Rb86
13	Zn65
14	Zr95
15	Ba131
16	Gd153

man-months were employed at the Lowie Museum, University of California, Berkeley, in order to obtain 100 obsidian samples associated with organic samples. The entire data set is presented in appendix 2.

In retrospect, the acquisition of an adequate data set remains as a definite limiting factor of hydration rate research. This problem is also illustrated by Clark's research program, where after many years of work, only a few data points were usable for a single hydration rate (Clark 1961a). It is recommended that in subsequent studies the data set should be acquired by the investigator through excavation of a series of test units or columnar sections within known habitation sites, ideally over a diverse region of space and time range. This recommendation amounts to the old maxim that " if one wants to do archaeological research, one must collect one's own data, since prior data is not sufficiently controlled or 'tailor-fit' to one's own specific needs today." Unfortunately, this maxim does not always favor the conservation of the archaeological record.

Selection Criteria

Three criteria were used in the selection of samples for hydration rate data. The criterion of association was that an included obsidian artifact had to have the same unit-level provenience as a given radiocarbon date. Generally, units were reported as 5 feet x 5 feet or meter square and levels were reported as 10 centimeters or 6 inches. The exact provenience of both the obsidian artifacts and radiocarbon date are presented in appendix 2. If the stratigraphy was complex, inverted, or disturbed as to cultural integrity or if the artifacts were "fired-burned" or other anomalies were observed, then the data were not included.

The second criterion was to obtain samples from diverse archaeological sites throughout California. This was to insure a sample of hydrated artifacts from all obsidian sources, which were used for the manufacture of chipped stone tools. The northern and northeastern sections of California were not sampled due to unavailability of samples. Fortunately, the Nightfire Island data set from Johnson's study (Johnson 1969) subsequently characterized by Dr. Al Waibel, Department of Geology, Portland State University, was made available by Professor C. Garth Sampson, Department of Anthropology, Southern Methodist University.

The third criterion was to obtain hydrated artifacts over a wide range of time during which a given obsidian source was utilized. In operationalizing these criteria, data from the Early through Late Horizons were sought for each obsidian source. This was to insure a sample of hydrated artifacts over the range of utilization of a given source. Generally, this range is less than 3,000 years, although the Mostin Site samples date to approximately 10,000 years ago (Ericson and Berger 1974).

The use of unit-level association between radiocarbon data and obsidian samples from regionally-diverse sites, sampled over a wide range of time, provided a means to control the data set. Although these three criteria provided the variable control and insured a sample over space-time variation, these restrictions acted to limit the quantity of data which could have been utilized. In retrospect, the type of criteria used in formulating the original hydration rates (Friedman and Smith 1960) perhaps would have generated workable source-specific rates for California, which would have been sufficient for hypothesis testing of chemical and physical relationships in later sections of this work.

Measurements for Hydration Rates

Notwithstanding the difficulties involved in assembling well-controlled samples, literally hundreds of obsidian artifacts and associated organic samples were processed for hydration rate determination. The procedures used in making these measurements merit description. The obsidian artifacts were chemically characterized as to source, and then hydration measurements were made. Organic samples were pretreated, converted into carbon dioxide, and then their radiocarbon content was determined. The data was subsequently stratified as to obsidian source and is presented in appendix 2.

The artifacts were chemically characterized by either X-ray fluorescence, short or long half-life radionuclide instrumental neutron activation analysis following the procedures described for the sources, and stepwise discriminate analysis, using the "packaged" programs now available in the Obsidian Hydration Dating Laboratory, UCLA. Each artifact was given a UCLA analysis number. This number and the provenience of each artifact which included the site, unit, level, collection, artifact number, and submitter of the samples along with the results of other analyses are in the UCLA Obsidian Logbook. The elementary analysis was recorded on computer-printed paper, inspected for errors, placed on punch cards, and entered with obsidian source data in a specific stepwise discriminate program. The computer output described the probability of membership to a particular source and the Mahalabinos D-squared statistic. The programs were designated as CC = Central California—short half-life; S = Southern California—short half-life; LC = All California—long half-life.

The Borax Lake and Mt. Konocti were distinguished by using ternary or tri-poled graphs of the elementary concentrations of zirconium, rubidium, and strontium determined by X-ray fluorescence analysis—a technique described by Jack and Carmichael (1969).

Each associated organic sample was given a UCLA radiocarbon number and and its provenience recorded. The results of radiocarbon dating will appear in the Radiocarbon journal (Berger and Libby n.d.).

The organic samples were pretreated with 1N NaOH to remove humic acid residues and 3N HCl to remove residual carbonates or bicarbonates. Root hairs and rootlets, if present, were removed by meticulous picking. The non-collagen portions of bones and teeth samples were dissolved by 6N HCl for periods up to one month. The collagen portion was used as the organic sample for radiocarbon dating. Shell samples were converted to carbon dioxide gas with 6N HCl. For the purposes of counting, a gas equivalent of a 4-gram pure carbon sample was needed for the 7.5 liter counter. The amount of sample was predetermined by original material: approximately 10 grams of charcoal, 15 grams of wood, 100 grams of shell, and up to 500 grams of bone. Occasionally, nitrogen microchemical analyses were performed on bone samples by Elek Microanalytical Laboratories. The percentage of carbon could be estimated by multiplying the nitrogen weight-percent by 3.5—this procedure was used to conserve the amount of bone. During any experiment it was very difficult to predict from the physical form of the collagen during acid extraction whether fine-colloidal particles or pseudomorphs of the original bone would result. This remains an interesting problem.

After drying, the pretreated samples were combusted in a closed oxygen environment. The combustion of the sample produced a gas mixture of carbon dioxide, water, some carbon monoxide, sulfur dioxide, chlorine, fluorine, bromine, iodine, ozone, nitrogen oxides, and radon. All components of the gas mixture except carbon dioxide and radon were removed by a system of interconnected catalysts and traps. Copper oxide at $600°C$ was used to oxidize any carbon monoxide. Three 1N silver nitrate solution traps were used to precipitate the halogens. Three chromic acid traps were used to remove sulfur dioxide and water. Manganese perchloride absorbed residual water. The carbon dioxide and radon were trapped within the combustion line by two liquid nitrogen cooled traps.

The low temperature of the cold traps and rapidity of the collection process entrapped oxygen, the precombustion carrier gas. Residual oxygen and electronegative impurities were removed from the carbon dioxide-radon mixture by passing the gas through $600°C$ pure copper, which was periodically reduced by hydrogen. The oxygen removal process was repeated from five to 15 times, in order to produce "countable" carbon dioxide. The "countable" carbon dioxide was stored in a high-pressure container for a period from four days up to one month to insure that the radon, which has a half-life of 3.8 days, would decay. At all times, the movement of the carbon dioxide was accomplished by alternatively sublimating the solid ("dry ice") to a gas received by a large copper cylinder or reversing the processes by freezing the gas with liquid nitrogen which surrounded a cold trap. The process of combusting and cleaning usually required three to four hours; occasionally, the process required up to eight hours.

The β-decay of the radiocarbon or Carbon-14 in the carbon dioxide gas was counted for a period of 1,000 minutes or 16 hours and 40 minutes in a 7.5 liter Geiger counter surrounded by an array of 13 Geiger-Müller anti-coincident counters. The decay of residual radon was counted and distinguished by a two-channel analyzer. The voltage or potential between the outside and inside counters was established and balanced prior to and during counting. Equation 1-1 described calculations involved in determining a radiocarbon date:

$$\text{Date (yrs.)} = 8,030 \ln^{-1} (X/y/T) \qquad [1-1]$$

where:

X = 95% oxalic acid standard (cpm)
y = Carbon 14 and Radon minus Radon minus Marble (cpm)
T = Counting time

The value of "modern" or 95% oxalic acid pre-1950 standard was approximately 40 counts per minute. The value of the "old" or marble or background standard was approximately 13 counts per minute. Thus, unknown samples ranged between 40 and 13 counts per minute depending on their age. The samples were counted two to three times to insure repeatability of the measurement.

The reported standard deviation "or ± factor" is actually a 2-sigma standard deviation based on the counting statistics rather than repeated measurements. The underlying reason behind this calculation is that two to three repeated results of 1,000 minute counting periods is not a significant sample to determine the range of the mean. The 2-sigma number is a function of age of the sample, counting period, and dilution. If the sample is not diluted, recent, and counted for a long period of time, the error can be minimally $\pm$ 40 years.

The calculated dates, reported in appendix 2, were determined using the tree-ring calibrated radiocarbon curve (Suess 1970). The radiocarbon dates are expressed in years "B.P." or years before 1950 A.D.

Hydration measurements were made on obsidian artifacts by Frank J. Findlow, Suzanne DeAtley, Victoria Bennett, and Gary Stickel at the UCLA Hydration Dating Laboratory. Multiple obsidian thin sections were cut from each specimen with a Felke Di-Met (Model 11-B) saw using a $3\frac{1}{2}$-inch diamond-charged, brass saw blade, ground optically-flat with silicon carbide #400 and aluminum oxide #95 grits on a flat plate, mounted on a petrographic slide with Lakeside plastic cement, ground on the other side to a thickness of 0.003 inch, covered with a glass slide using Canada balsam, and labeled with a diamond stylus. The obsidian hydration layers were measured at 537.5X magnification, using a Leitz micrometer eyepiece mounted in an American Optical petrographic microscope. Measurements were taken at one to four different points with five readings made per point. The results of these observations were averaged for each thin section. The pooled standard deviation for each thin section was 0.2 micron which included the three components of variability, namely hydration rim variability, instrument error, and observer error (Aiello 1969). The hydration measurements are reported in appendix 2.

EMPIRICAL OBSIDIAN HYDRATION RATES FOR CALIFORNIA

More than determining source-specific obsidian hydration rates for California, a project important in its own right, the controlled data set in table 1-8 provided a means to examine a number of important questions which were still unresolved by prior research. In a sense, the California data and hydration rates transcend their regional archaeological utility to give us a clearer understanding of the complex obsidian hydration phenomenon for a worldwide perspective.

Some important considerations are the determination of source-specific hydration rates and their statistical confidence intervals for direct chronological application in California, an evaluation of the accuracy gained by source specification, and an outline of procedures for simplifying these identifications. Finally, there is the evaluation of the general hydration models proposed in the literature as to the degree of statistical fit.

Source-Specific Hydration Rates

Obsidian hydration measurements and associated radiocarbon data stratified by source are presented in table 1-8, which is a condensed version of the same data presented in appendix 2. These data were used to define a set of mathematical descriptions of interrelationships between time of formation and the hydration layer thickness. In particular, these mathematical descriptions are called the source-specific empirical hydration rate equations. The general mathematical structure of this variation is described in equation 1-2 and logarithmically in equation 1-3:

$$T = ax^b \qquad [1\text{-}2]$$

where:

T = time in years before present
x = hydration measurement in microns
a and b are constants

$$\log T = \log a + b \log X \qquad [1\text{-}3]$$

Estimates for the parameters a + b were obtained using least squares linear regression analysis of the data in table 1-8. Linear regression analysis of the data, stratified by source, was performed by the computer

Table 1-8

DATA FOR SOURCE-SPECIFIC OBSIDIAN HYDRATION RATE CURVES
CONDENSED FROM APPENDIX 3

Source	N	Hydration	Years B.P.	Temperature (C^o)
Annadel	6	2.50	570	27.5
		2.20	490	22.5
		2.65	3,575	19.0
		2.70	3,250	19.0
		4.30	4,090	22.0
		6.18	925	20.0
Bodie Hills	7	3.70	1,590	20.0
		4.80	2,775	24.0
		5.50	4,850	24.0
		2.03	270	18.0
		3.90	3,150	22.0
		2.40	1,550	22.5
		2.35	1,250	22.5
Borax Lake	10	2.30	270	18.0
		2.60	545	22.5
		7.40	7,750	25.0
		6.28	10,260	25.0
		4.40	1,890	23.0
		3.50	440	23.0
		4.10	4,850	24.0
		3.77	4,070	23.0
		3.21	2,080	23.0
		0.80	670	23.0
Buck Mt.	10	2.25	2,130	15.0
		3.10	2,400	15.0
		3.10	2,130	15.0
		3.40	3,960	15.0
		3.20	3,960	15.0
		4.10	4,920	15.0
		4.00	5,610	15.0
		4.20	4,920	15.0
		4.55	6,550	15.0
		4.56	7,075	15.0
Casa Diablo	15	6.13	930	24.0
		5.70	1,080	22.0
		6.76	510	24.0
		1.07	525	19.0
		3.19	525	19.0
		1.59	290	19.0
		3.66	620	19.0
		3.30	670	19.0
		3.30	2,050	19.0
		5.20	1,590	20.0 (contd.)

Table 1-8 (contd.)

Source	N	Hydration	Years B.P.	Effective Temperature (C°)
Casa Diablo (continued)	15	2.90	850	24.5
		5.40	1,165	22.0
		4.10	1,890	23.0
		2.85	1,250	22.5
		4.66	450	16.0
Coso	14	12.27	1,870	25.0
		4.95	2,080	29.0
		2.00	1,670	29.0
		0.70	280	24.5
		4.23	1,165	22.0
		6.33	1,275	25.0
		8.03	2,320	23.0
		8.04	3,160	23.0
		8.40	3,975	23.0
		8.16	4,010	23.0
		8.48	4,440	23.0
		4.62	1,590	29.0
		4.60	1,920	29.0
		3.60	740	24.0
Modoc Glass Mt.	11	2.45	2,130	15.0
		2.69	2,400	15.0
		2.70	2,130	15.0
		3.50	3,960	15.0
		3.70	3,960	15.0
		3.64	4,920	15.0
		4.20	5,610	15.0
		4.58	4,920	15.0
		4.22	6,550	15.0
		4.43	7,075	15.0
		2.35	1,500	15.0
Mono Craters	4	5.98	510	24.0
		6.45	930	24.0
		8.40	2,320	23.0
		7.70	3,070	23.0
Mono Glass Mt.	6	4.23	925	20.0
		4.40	1,590	20.0
		5.90	3,375	24.0
		4.70	2,050	22.0
		5.80	2,080	29.0
		2.50	1,550	22.5
Mt. Konocti	4	4.60	10,260	25.0
		4.08	9,040	25.0
		4.40	7,750	25.0
		4.05	4,850	24.0
Napa Glass Mt.	23	3.00	3,575	24.0
			3.13	3,250	19.0

(contd.)

Table 1-8 (contd.)

Source	N	Hydration	Years B.P.	Effective Temperature (C°)
Napa Glass Mt.		2.30	1,890	23.0
(continued)		3.80	3,150	19.0
		2.36	500	22.0
		2.60	450	24.0
		3.86	3,150	24.0
		2.70	1,250	22.5
		2.87	1,050	22.0
		2.17	650	22.0
		3.00	3,375	24.0
		3.40	2,000	22.0
		2.36	540	22.0
		2.00	1,050	22.0
		2.30	270	18.0
		2.45	550	22.5
		2.22	570	22.5
		2.36	490	22.0
		2.55	490	22.0
		3.90	3,575	19.0
		2.60	4,850	23.0
		4.45	2,830	24.0
		1.69	1,080	24.0
Obsidian Butte	2	2.77	840	24.5
		2.70	280	24.5
Mt. Hicks	3	5.40	925	20.0
		6.77	1,590	20.0
		2.20	1,590	22.5
Pine Grove Hills	2	4.30	925	20.0
		4.80	4,850	24.0

program, BMDP1R (Jackson and Douglas 1975). The results of the regression are presented in table 1-9.

The process of formulating the empirical hydration rates requires some qualifications. The empirical hydration rates are mathematical descriptions of the data, i.e., they are the best statistical fit to the data which is now available as compared to fitting the data to the best physical model. The best statistical fit is data dependent in that it assumes that the data are accurate and that uncontrolled variables have little effect on the variables under study. On the other hand, one can argue, as did Friedman and Smith in 1960, that the hydration process follows and agrees with a physical model such as Fick's Second Law of Diffusion with some of the initial conditions being concentration-dependence of the diffusing water into a semi-infinite planar solid. A number of other physical models and empirical equations have been proposed as discussed in Ericson and Berger (1976). Each claims a certain parity between statistical fit versus a physical model. In the

Table 1-9

EMPIRICAL SOURCE-SPECIFIC OBSIDIAN HYDRATION DATING
EQUATION USING THE MATHEMATICAL MODEL:
Log T = Log a + b Log X (Equation 1-4)

Source	N	Log a	b	Pearson's R
All	108	2.6441	1.092	0.5201
Annadel	5	1.9602	2.837	0.6823
Bodie Hills	7	2.1505	2.059	0.8535
Borax Lake	9	2.5255	1.332	0.6235
Casa Diablo	112	2.5868	0.671	0.5099
Coso	13	2.6247	0.932	0.8602
Mono Craters	4	-1.0376	4.910	0.9284
Mono Glass Mt.	6	2.7974	0.705	0.5151
Mt. Konocti	4	1.7234	3.426	0.6421
Napa Glass Mt.	20	1.6317	3.287	0.7656
Obsidian Butte	2	---	---	---
Mt. Hicks	3	3.2128	-0.147	0.2865
Pine Grove Hills	2	---	---	---
Buck Mt.	10	2.5918	1.830	0.9035
Modoc Glass Mt.	11	2.4928	2.015	0.9607

following sections, these models and equations are examined relative to the data set present in table 1-8.

In formulating the hydration rates, the temperature variable cannot be neglected as shown originally by Friedman and Smith (1960). The next section is an attempt to incorporate temperature as a variable in developing hydration rates.

Incorporation of Temperature as a Variable in Empirical Hydration Rates

Temperature as a variable is already incorporated in the Fick's diffusion equations as shown in equations 1-4 and 1-5:

$$X^2 = kt \qquad [1\text{-}4]$$

where:

X = hydration thickness
k = diffusion coefficient
t = time

$$k = A \exp(-E/RT) \qquad [1\text{-}5]$$

where:

k = diffusion coefficient
A = constant
E = activation energy
R = gas constant
T = temperature

However, among the empirical rates so far determined, temperature is omitted as a variable. Since the hydration process is a function of at least two independent variables, temperature should be incorporated as a variable. This section attempts to incorporate temperature as a variable and to formulate a simple procedure by which archaeologists can apply it to their own data.

Temperature, or what is better termed an effective temperature variable, is quite important. Fortunately, meteorologists have documented the mean annual air temperature and average range of variation as a function of time which provides a usable record. However, most natural reactions like the hydration process proceed as an integrated mean temperature. In these endothermic (heat-absorbing) reactions, the <u>rate of change</u> of the reaction increases as a function of temperature. Therefore, the fluctuations of temperature around the mean annual air temperature have to be considered, i.e., temperatures above the mean are increasingly more effective. In discussing this problem Lee (1969:Eq. 12) proposes an equation, equation 1-6, which allows an estimation of what we will call the <u>natural effective temperature:</u>

$$T_a = -1.2316 + 1.0645 T_e - 0.1607 R_T \qquad [1\text{-}6]$$

where:

T_a = Mean Annual Air Temperature
T_e = Natural Effective Temperature
R_T = Temperature Range of T_a

In this study, the natural effective temperature of each data point on table 1-8 was estimated. The mean annual air temperature varies extensively in California so as to include subarctic to even tropical thermal regimes. The operation of prehistoric exchange systems where obsidian was an exchange item provided a mechanism by which obsidian was distributed throughout all temperature regimes. Thus, the source-specific obsidian data may represent source samples from many thermal regions. The easiest solution to this problem would be to determine the mean annual air temperature for each archaeological site in this study. Unfortunately, a mean annual air temperature map of California could not be located. Thus, for this and future application, a natural effective temperature map had to be produced, shown in plate 1-3.

The natural effective temperature map of California was produced by using the temperature data presented by Felton (1965). The mean annual air temperature, the temperature range, determined by subtracting the average maximum and minimum, and geographical locations of 120 weather stations were used as data for the production of the map (plate 1-3). The mean air temperature and the range were converted to degrees Centigrade and in turn to a natural effective temperature using equation 1-6. The name of the weather station, geographical location, mean annual and natural effective temperatures are presented in table 1-10. These values were entered as data in SYMAP computer program using three passes to produce a map with $2^\circ C$ temperature intervals, shown in plate 1-3. The natural effective temperature of each archaeological site was determined using this map, plate 1-3, and tabulated in table 1-8.

Plate 1-3 Natural Effective Temperature Map of California. There are ten 2.03°C temperature contour intervals on this map. The full range of temperatures is 11.7 to 32.0°C. The minimum and maximum of each zone are the following: Zone 1, 11.70-13.73; Zone 2, 13.73-15.76; Zone 3, 15.76-17.79; Zone 4, 17.79-19.82; Zone 5, 19.82-21.85; Zone 6, 21.85-23.88; Zone 7, 23.88-25.91; Zone 8, 25.91-27.94; Zone 9, 27.94-29.97; Zone 10, 29.97-32.00 in degrees Centigrade.

It is important to mention several types of improvements in estimating the natural effective temperature which go somewhat beyond the scope of this dissertation. On-site temperature measurement as a function of depth would be a definite improvement over the use of the natural effective temperature map. The natural effective temperature map considers only the air surface temperature, not the surface soil temperature or soil temperature as a function of depth. An improvement of the utility of the map would be to correct the surface value as a function of soil depth. Apparently, general empirical equations which link surface air temperatures as a function of depth have not yet been developed. These equations are lacking due to the number of variables involved in soil temperature. A diagram of the variation of the natural effective temperature relative to soil depth and mean annual air temperature is shown in plate 1-4. Thus, although there are a number of improvements in estimating the natural effective temperature of the environment for a given obsidian sample which are beyond the scope of this project, the natural effective temperature as determined here is used as a second independent variable.

In this further analysis, the hydration process was considered to be a function of at least two independent variables, namely, time and temperature as shown in equations 1-7 and 1-8:

$$T = cx^d t^e \qquad [1\text{-}7]$$

where:

T = time in years before present
x = hydration microns
t = natural effective temperature (absolute)
c, d, and e = constants

$$\log T = \log c + d \log x + e \log t \qquad [1\text{-}8]$$

(same symbols)

Multiple linear regression of the hydration, temporal, and temperature data was performed by the computer program BMDPIR. The results of the regression are not presented as there was little improvement of Pearson's R by incorporating temperature. In the next section, the several physical models and empirical equations are evaluated for their exactness of fit.

An Evaluation of Physical Models and Empirical Equations Describing the Hydration Process

In the section entitled "Source-Specific Hydration Rates," it was pointed out that researchers have in the past fit their data to either statistically-defined mathematical models or physical models, such as the diffusion model of Friedman and Smith (1960) or the autocatalytic model, suggested by Ericson (1975). The particular usefulness of the California data as a well-controlled data set would be to evaluate each of the obsidian hydration dating models already published in the literature. Such a comparative study would go far in bringing a greater understanding of the type of variability involved in the

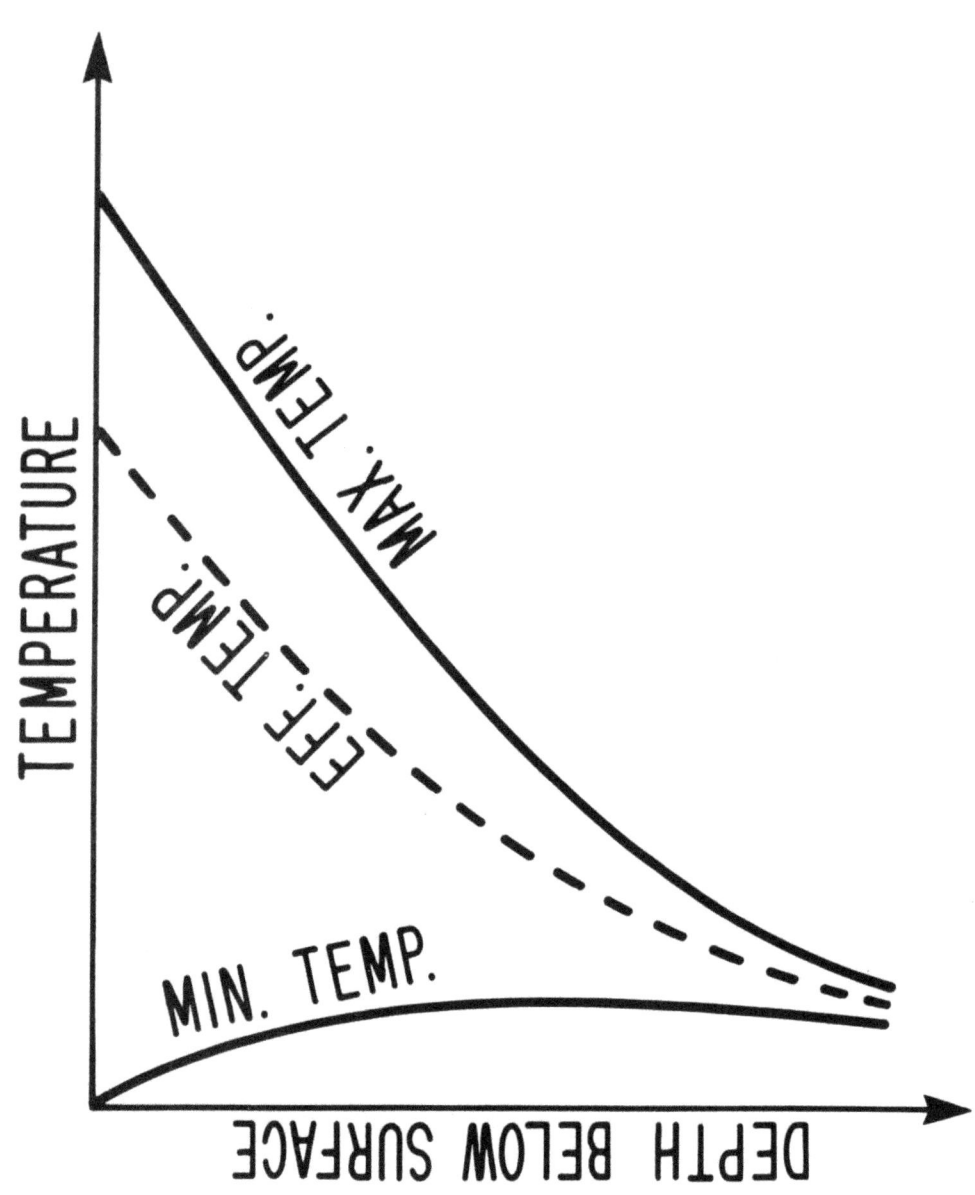

Plate 1-4. Diagram of Natural Effective Temperature as a Function of Depth in Soil.

Table 1-10

GEOGRAPHICAL LOCATION, MEAN ANNUAL AIR TEMPERATURE,
AND NATURAL EFFECTIVE TEMPERATURE FOR 120 WEATHER
STATIONS IN CALIFORNIA
(Modified after Felton 1965)

Station	Lat. (N)	Long. (E)	Temp. (°C)	Natural Effective Temperature
Alpine	32 50	116 46	23.5	25.8
Auburn	38 54	121 04	22.9	24.8
Avalon Pleasure Pier	33 21	118 19	19.7	20.8
Bakersfield WSO AP	35 25	119 03	26.0	28.0
Barstow	34 54	117 02	26.7	29.0
Beaumont	33 56	116 58	23.5	25.6
Bishop WSO AP	37 22	118 22	23.0	25.8
Blue Canyon WSO AP	29 17	120 42	14.5	16.2
Blythe	33 37	114 36	31.2	33.4
Brawley 2 SW	32 57	115 33	31.8	33.8
Bridgeport	38 15	119 14	15.7	18.7
Burbank Valley	34 11	118 21	24.2	26.1
Calexico 2 NE	32 41	115 28	30.2	32.1
Chico University Farm	39 42	121 49	24.2	26.3
Coalinga	36 09	120 21	26.1	28.3
Colusa 2 SSW	39 12	122 01	23.4	25.3
Crescent City	41 96	124 12	16.7	18.3
Davis 2 WSW EXP Farm	38 32	121 46	23.8	26.1
East Park Reservoir	39 22	122 31	23.6	26.0
El Centrol 2 SSW	32 46	115 34	32.0	34.0
Escondido	33 07	117 05	24.5	26.7
Eureka WSO C1	40 46	124 10	14.1	15.3
Forest Glen	40 23	123 20	19.4	22.1
Fort Bragg	39 27	123 48	16.2	17.7
Fresno WSO AP	36 46	119 43	24.6	26.5
Gem Lake	37 45	119 08	11.7	14.1
Grass Valley No. 2	39 13	121 04	21.4	23.2
Half Moon Bay	37 28	122 27	16.7	18.1
Happy Camp Ranger Station	41 48	123 22	16.7	19.4
Idyllwild Fire Department	33 45	116 43	19.2	21.8
Independence	36 48	118 12	22.4	24.6
Indio U.S. Date Garden	33 44	116 15	31.4	33.3
Kentfield	33 57	122 33	21.6	23.8
La Mesa	32 46	117 01	23.9	25.7
Lake Arrowhead	34 15	117 11	18.3	20.5
Long Beach WSO AP	33 49	118 09	22.2	23.6
Los Angeles Civic Center	34 03	118 14	22.8	24.2
Los Angeles WSO AP	33 56	118 24	20.7	22.0
Los Banos	37 03	120 52	25.1	27.3
Los Gatos	37 14	121 58	21.4	23.3 (contd.)

Table 1-10 (contd.)

Station	Lat. (N)	Long. (E)	Temp. (°C)	Natural Effective Temperature
Merced Fire Station 2	37 18	120 29	24.3	26.4
Mineral	40 21	121 36	15.2	17.7
Monterey	36 36	121 54	18.7	20.5
Mt. Hamilton	37 20	121 39	15.6	17.0
Mt. Shasta WSO CI	41 19	122 19	17.0	19.4
Mt. Tamaplais 2 SW	37 54	122 36	16.2	17.4
Mt. Wilson 2	34 14	118 04	18.6	30.5
Needles	34 50	114 36	29.6	22.1
Newark	37 31	122 02	20.2	21.2
Newport Beach Harbor	33 36	117 53	20.0	20.3
Oakland WSO AP	37 44	122 12	18.8	21.1
Oceanside	33 12	117.23	19.9	28.1
Ojai	34 27	119 15	25.6	26.2
Orland	39 45	122 12	24.3	26.7
Palmdale	34 35	118 06	24.5	33.1
Palm Springs	33 50	116 30	31.1	22.0
Palo Alto	37 27	122 08	20.2	20.3
Palomar Mt. Observatory	33 21	116 52	18.6	23.6
Paradise	39 45	121 37	21.9	31.5
Parker Reservoir	34 17	114 10	30.1	26.2
Pasadena	34 09	118 09	24.3	25.6
Paso Robles	35 38	120 41	24.5	23.5
Placerville	38 44	120 48	21.1	16.5
Point Piedras Blancas	35 40	121 17	15.3	27.3
Pomona Cal Poly	34 04	117 49	25.0	27.8
Porterville	36 04	119 01	25.7	20.8
Portola	39 48	120 28	17.6	24.4
Priest Valley	36 11	120 42	21.6	27.1
Ramona Fire Department	33 03	116 52	24.8	25.3
Red Bluff WSO AP	40 09	122 15	23.6	27.0
Redland	34 03	117 11	25.0	28.5
Riverside Fire Station # 3	33 57	117 23	26.2	23.8
Sacramento WSO C1	38 35	121 30	22.1	25.5
Saint Helena	38 30	122 28	23.2	21.6
Salinas 3 E	36 40	121 36	19.8	29.1
San Bernardino Co. Hosp.	34 08	117 16	26.8	21.3
San Diego WSO AP	32 44	117 10	20.2	18.0
San Francisco WSO C1	37 47	122 25	16.9	22.9
San Jose	37 21	121 54	21.1	23.0
San Luis Obispo Poly	35 18	120 40	21.2	23.3
Santa Ana Fire Station	33 45	117 52	24.3	23.4
Santa Barbara	34 25	119 41	21.7	22.7
Santa Cruz	36 59	122 01	20.7	22.4
Santa Monica Pier	34 00	118 30	21.1	22.4

(contd.)

Table 1-10 (contd.)

Station	Lat. (N)	Long. (E)	Temp. (°C)	Natural Effective Temperature
Santa Paula	34 21	119 05	23.7	25.8
Santa Rosa	38 27	122 42	21.9	24.1
Soda Springs	39 19	120 22	12.7	15.5
Sonora RS	37 59	120 23	22.8	24.8
Stockton WSO AP	37 54	121 15	22.3	24.1
Susanville 1 WNW	40 26	120 40	17.4	19.8
Tehachapi	35 08	118 27	19.9	22.2
Tejon Rancho	35 02	118 45	23.6	25.1
Three Rivers Edison PH1	36 28	118 52	25.4	27.7
Truckee Ranger Station	39 20	120 11	15.1	17.9
Tule Lake	41 58	121 28	16.5	19.2
Ukiah	39 09	123 12	23.2	25.6
Upper Lake 7 W	39 11	123 02	22.6	25.0
Victorville Pump Plant	34 32	117 18	25.3	28.0
Visalia	36 20	119 18	25.8	27.9
Walnut Creek 2 ENE	37 54	122 01	22.1	24.1
Weaverville Ranger Sta.	40 44	122 56	21.3	24.0
Woodland	38 41	121 48	24.0	26.1
Yosemite Park Headquarters	37 45	119 35	20.2	22.8
So. Entrance Yosemite	37 30	119 38	18.1	20.7
Yreka	41 43	122 38	19.4	21.9

hydration process, if not a resolution to particular issues or even, indeed, the debate itself. Such an evaluation can be conducted by observing the degree of internal consistency or statistical fit of each model to the data for all the sources. If a particular model can be considered general, a convergence of fit toward one particular model should be observed as successively more controls are imposed on the data. For example, it has already been demonstrated that source specificity does appear to be an important stratification of the data (compare Pearson's R for "all data" equation and the source-specific equations).

Five models were selected from the research literature on the hydration process:

1. <u>Linear Rate Model</u>—Meighan, Foote, and Aiello (1969) found that the expression used by Friedman and Smith (1960) did not satisfactorily describe the relationship between hydration and time for the Morret site in West Mexico. They suggested a linear hydration equation for West Mexico, $X = Dt$, where $D = 3.85$ microns/10^3 years based on the relationship of 16 radiocarbon dates and 115 obsidian hydration measurements of artifacts.

   Morgenstein and Riley (1973) found that the expression used by Friedman and Smith did not satisfactorily describe hydration phenomena.

They calculated a linear hydration rate for Hawaiian basaltic glasses based on $X = Dt$ where $D = 11.77$ microns$^2/10^3$ years. Furthermore, they suggest that the reaction of water with glass does not slow down with increasing depth of hydration but rather remains constant, which accounts for the linear rate.

2. <u>California Rate Model</u>—The California rate model was proposed by Clark. Clark (1961a, 1961b, 1964) found that the expression $X = Dt^{\frac{1}{2}}$ did not satisfactorily describe the relationship between hydration and time. Thus, he suggested a central California hydration equation, $X = Dt\ 3/4$, based on the hydration of obsidian artifacts and radiocarbon dates. These results are also reported and discussed by Friedman, Smith, and Clark (1963). Later, Johnson (1969) presented an obsidian hydration rate for the Klamath Basin of California and Oregon based on the relationship between 10 radiocarbon-dated levels and 107 obsidian artifacts. He suggested the rate of 3.54 microns$^2/10^3$ radiocarbon years for this area. The Johnson rate is not considered as an original model in this analysis.

3. <u>Diffusion Rate Model</u>—The diffusion rate model was originally proposed by Friedman and Smith (1960). In this original paper, hydration rates for each of the Earth's major climatic zones were suggested. The thermodynamic nature of the hydration process was preliminarily examined by Friedman, Smith, and Long (1966).

   Katsui and Kondo (1965) presented a similar (diffusion) hydration rate for Hokkaidō, Japan, based on data from six radiocarbon-dated levels with obsidian artifacts distributed from 1,000 to 15,000 years. The rate was 1.6–2.0 microns$^2/10^3$ years which is in the interval between the rates 4.5 microns$^2/10^3$ years and 0.82 microns$^2/10^3$ years established for Temperature Zone 2 and the subarctic Zone, respectively, established after Friedman and Smith (1960).

4. <u>Cubic Rate Model</u>—Kimberlin (1971) using the hydration results of a group of chemically related obsidian artifacts from the Morret site, Colima, Mexico, suggested the hydration equation $X = Dt^3$. Also, he illustrated the influence of the initial concentration of water on the rate of hydration.

5. <u>Parabolic-Rate Model</u>—The parabolic rate was suggested by empirical fit to obsidian hydration rate data of the Southwest United States by Findlow et al. (1975). Ericson (1975) attempted to link this mathematical model with the physical process of autocatalytism where water both diffused and reacted during the hydration process.

6. <u>Square Root Rate Model</u>—A sixth mathematical model, the square root rate model, was established here such that the exponents of all the models would vary from 0.5 to 3.0 following the general equation form developed in equation 1-2.

Linear regression analysis of the source-specific data was performed by the computer program BMDP1R using each of the above equations. The results of the regression are presented in table 1-11. As already demonstrated, it appears that source-specificity is an important operation of the empirical curve-fitting process and/or physical model testing. This is illustrated by the range of variability of the rates for any given mathematical model. The degree of fit of each case is demonstrated by Pearson's regression correlation coefficients which are tabulated in table 1-12. The diffusion model appears to be the best model in five of 14 cases. This data does not conclusively resolve the standing question of model.

The diffusion model can be maintained as the general hydration model as long as the obsidian hydration data are stratified by source. This is the main thesis of the monograph; namely, that source-specific rates increase the accuracy of the dating technique.

CONCLUSIONS

After extensive testing, it does appear that the original diffusion model (Friedman and Smith 1960) provides the best physical model and a general mathematical description of the hydration process for rhyolitic obsidian. However, this study has definitely demonstrated the need for source-specific hydration rates for increased accuracy of the dating technique. In addition, the measurement of the hydration (thermal) environment perhaps will be a further improvement. These results suggest that the variety of descriptive models in the literature may be the result of inadequate control of temporal association, source specification, or hydration (thermal) environment.

For California, where many different obsidian sources are in the archaeological record, source specification has required the field identification of sources and the collection of samples which were characterized by both chemical and multivariate analysis. As a result, it is now possible on a routine basis to chemically characterize obsidian artifacts without knowing their regional provenience within California. This can be accomplished by using long half-life radionuclide instrumental neutron activation analysis of artifacts and the "packaged" program now available at the Obsidian Hydration Dating Laboratory, Department of Anthropology, UCLA.

As formulated herein, the application of the long half-life radionuclide INAA technique of chemical characterization provides the greatest precision and an objective means to characterize obsidian artifacts leading to the routinization of laboratory procedure. Nevertheless, depending upon the particular research strategy and the facilities available, two other procedures can be used for source identification. The rapid-scan X-ray fluorescence technique (Jack and Carmichael 1969) or the short half-life radionuclide INAA technique, developed herein, are useful if the number of sources, prehistorically utilized at a particular time or place, is known to be small or chemically distinctive.

Selection criteria in hydration rate determination have been shown to be very important. In this study, three sampling criteria were used in selecting the data, namely, unit-level association between artifacts and radiocarbon samples, from diverse sites, and over the widest time span. Although the type of data control used in formulating the original hydration rates (Friedman and Smith 1960) would have generated workable source-specific rates for California, which would have been sufficient for hypothesis testing of chemical and physical relationships in later sections of this work, it would not have resulted in any great improvement in the accuracy of the dating techniques. The empirical source-specific obsidian hydration dating rates (diffusion model) presented in table 1-12 are now available for use and evaluation. Finally, as a result of the above research, a number of new radiocarbon dates are also available for California archaeologists.

Table 1-11

RATE CONSTANTS OF SOME OBSIDIAN HYDRATION MODELS

Model: Type: Reference: Equation:		1 Square Root (none) $T = aX^{\frac{1}{2}}$	2 Linear (Meighan, Foote, & Aiello 1968) $T = bx$	3 California (Clark 1964) $T = CX^{1.33}$	4 Diffusion (Friedman & Smith 1960) $T = dX^2$	5 Cubic (Kimberlin (1976) $T = eX^3$	6 Parabolic (Findlow et al. 1975) $T = f(X^2-X)$
Source	N	A	B	C	D	E	F
Annadel	6	1148.016	537	303.223	84.297	10.902	94.479
Bodie Hills	7	1273.717	670	420.687	152.813	30.611	193.461
Borax Lake	10	1953.041	979	582.064	180.606	36.642	215.849
Buck Mt.	10	2356.052	1234	794.196	316.093	76.115	420.117
Casa Diablo	15	487.280	111	127.806	39.532	6.432	47.126
Coso	14	946.151	344	167.056	34.980	2.855	38.372
Modoc Glass Mt.	11	2284.323	1220	792.06	317.757	76.600	423.225
Mono Craters	4	660.777	252	123.112	35.002	4.628	40.527
Mono Glass Mt.	6	920.446	421	246.813	81.270	14.786	99.822
Mt. Konocti	4	3869.361	1873	1159.318	436.310	100.752	568.134
Napa Glass Mt.	23	1112.325	670	466.960	211.602	57.183	298.541
Obsidian Butte	2	--	--	--	--	--	--
Mt. Hicks	3	597.466	240	128.928	36.091	5.409	42.042
Pine Grove Hills	2	--	--	--	--	--	--

Table 1-12

DEGREE OF FIT OF THE HYDRATION RATE MODELS MEASURED BY PEARSON'S R

Source	N	1	2	3	4	5	6
Annadel	6	0.808	0.755	0.702	0.580	0.425	0.526
Bodie Hills	7	0.916	0.958	0.971	0.976	0.955	0.970
Borax Lake	10	0.826	0.897	0.923	0.935	0.899	0.931
Buck Mt.	10	0.961	0.979	0.987	0.993	0.886	0.991
Casa Diablo	15	0.885	0.858	0.831	0.771	0.685	0.743
Coso	14	0.927	0.918	0.895	0.820	0.680	0.802
Modoc Glass Mt.	11	0.952	0.975	0.984	0.986	0.974	0.985
Mono Craters	4	0.884	0.908	0.921	0.941	0.954	0.944
Mono Glass Mt.	6	0.953	0.961	0.962	0.957	0.942	0.952
Mt. Konocti	4	0.974	0.977	0.979	0.981	0.983	0.982
Napa Glass Mt.	23	0.836	0.863	0.874	0.877	0.843	0.868
Obsidian Butte	2	---	---	---	---	---	---
Mt. Hicks	3	0.942	0.892	0.862	0.816	0.783	0.800
Pine Grove Hills	2	---	---	---	---	---	---

Chapter 2

HYDROTHERMALLY-INDUCED HYDRATION EXPERIMENTS

INTRODUCTION

The results of source-specific hydration rate determination in chapter 1 illustrate a number of difficulties in undertaking the research. Firstly, rates for all sources were not determined, due to sampling problems. Secondly, rate determination controlled only one environmental variable, namely temperature. It is quite possible that other, as yet unidentified, environmental variables may influence the accuracy of the dating technique. Thirdly, if the research costs of rate determination of California sources were projected for undertaking a worldwide research project, the costs would be prohibitive. In order to overcome these problems, a program of controlled induced hydration experiments was conducted over the past five years. The results of these experiments are presented in this chapter and compared with results obtained by archaeological data control.

The purpose of induced obsidian hydration experiments is to produce hydration bands under controlled laboratory conditions. At elevated temperatures and pressures or hydrothermal conditions, the hydration process can be activated quite rapidly. If the induced process of hydration is shown to be the same as the natural process, then rate determination can be accomplished much more simply in the laboratory. For example, it would be easy to establish hydration rates for most sources on a worldwide basis within one year. Nevertheless, it still remains to be demonstrated in this chapter that there is a concurrence between the natural and induced processes.

THE EXPERIMENTS

Beginning in 1972 and continuing until 1977, 17 induced hydration experiments were conducted. A record of the experiments and their parameters is presented in table 2-1. The success of the experiments in inducing observable hydration bands was not very good. The reasons behind the success or failure of a given experiment are not well understood—it does appear that there is a thermal discontinuity at approximately $190^{\circ}C$ above which the hydration band does not appear. This may be due to the annealing of the stress birefringence of hydration band upon cooling. Alternatively, the discontinuity may be due to a changeover in the mechanism of hydration. The actual reason for these anomalous results is of great interest but beyond the scope of this research project.

The procedures for hydrothermal induction experiments began in the mid-1950s with the extensive work of Bowen and others in studying crystalline phase transitions within rocks. This research provided the basis for much of

Table 2-1

EXPERIMENTAL CONDITIONS OF INDUCED
HYDRATION EXPERIMENTS

Exp. No.			Time (Days)	Temp. (°C)	Phase	Pressure (psi)	Notes
1	1/14/72	1/24/72	10	200	1	215	No hydration observed
2	2/10/72	2/22/72	10	258	2	675	No hydration observed
3	10/8/74	10/12/74	4	163	2	97	Degassed end of exp., table 2-3
4	12/18/75	12/22/75	4	200	2	225	See table 2-5
5	12/18/75	12/29/75	11	200	2	225	See table 2-5
6	12/18/75	1/5/76	18	200	2	225	See table 2-5
7	12/19/75	1/12/76	24	163	2	97	See table 2-3
8	12/22/75	2/5/76	25	150	2	69	See table 2-2
9	12/22/75	1/6/76	15	163	2	97	See table 2-3
10	12/29/75	1/23/75	20	150	2	69	See table 2-2
11	1/5/76	1/20/76	15	150	2	69	See table 2-2
12	12/13/76	12/16/76	3	192	2	190	No hydration observed
13	12/13/76	12/20/76	7	192	2	190	No hydration observed
14	12/13/76	12/24/76	11	192	2	190	No hydration observed
15	2/24/77	3/1/77	3	172	2	121	See table 2-4
16	2/24/77	3/7/77	13 3/4	172	2	121	See table 2-4
17	2/24/77	3/24/77	28	172	2	121	See table 2-4

experimental petrology and crystal chemistry. Through hydrothermal induction as a simulation process, the physical and chemical conditions underlying the states of matter could be ascertained.

The simulation of obsidian hydration began with experiments by Nasedkin in 1964. In 1972 an attempt was made to reproduce the results of this experimenter using the hydrothermal conditions set forth (Nasedkin 1964). A hydrothermal system developed by Professor George Kennedy of UCLA was used for the first two experiments. The stainless steel reaction vessel was

Table 2-2

RESULTS OF 150°C INDUCED HYDRATION EXPERIMENTS

Source	(Exp. 11) 15 Days	(Exp. 10) 30 Days	(Exp. 8) 45 Days	$K_1 \times 10^6$ $\mu^2/10^3 \text{yr}$
Annadel	6.1	5.4	---	0.630
Benton Valley, W. Queen	---	4.8	6.5	0.311
Bodie Hills	---	---	---	---
Borax Lake, Area 1	---	8.2	7.3	0.625
Borax Lake, Area 2	---	---	---	---
Buck Mt.	3.9	5.7	4.9	0.370
Casa Diablo, Section 29	---	---	---	---
Coso	3.9	---	6.3	0.346
Cowhead Lake, Section 34	4.0	6.1	6.6	0.398
Cowhead Lake, 8-Mile Creek, No.	3.7	5.1	6.6	0.334
Fish Springs	---	5.3	6.2	0.326
Inyo Craters, Hill 8160	---	---	---	---
Inyo Craters, Hill 8491	---	---	---	---
Inyo Craters, Hill 8611N	---	---	---	---
Jawbone Canyon	3.8	---	6.0	0.322
Modoc Glass Mt., Dacite Flow	---	---	---	---
Modoc Glass Mt., Rhyolite	3.5	4.3	4.0	0.218
Modoc Glass Mt., Little Glass Mt.	3.75	4.9	6.3	0.319
Modoc Glass Mt., Med. L. Glass Flow	---	---	---	---
Monache Meadows	---	---	---	---
Mono Craters, North Crater	4.9	6.4	---	0.541
Mono Craters, Northwest Coulee	4.9	7.1	6.7	0.364
Mono Glass Mt., Section 17	---	---	---	---
Mt. Konocti	---	5.1	5.3	0.273
Napa Glass Mt., West	---	---	---	---
Napa Glass Mt., East	4.1	6.0	---	0.424
Napa Glass Mt., South	3.8	5.0	6.3	0.326
Napa Glass Mt., E. Dago Valley	4.3	---	6.5	0.396
Napa Glass Mt., W. Dago Valley	4.6	5.4	---	0.435
Napa Glass Mt., Hill 450+	3.8	6.3	5.8	0.369
Obsidian Butte, Northwest	---	---	---	---
Sugarhill, Section 21	---	---	---	---
Duck Flat, Nev.	3.6	---	6.5	0.329
Fletcher, Nev.	---	---	---	---
Mt. Hicks, Nev.	---	6.3	7.6	0.476
Long Valley, Nev.	4.7	5.1	7.1	0.559
Pine Grove Hills, Nev.	4.5/3.6	6.3/6.1	---/6.2	0.488/0.360
Beatty's Butte, Ore.	7.1	---	4.8	0.707
Glass Butte, Ore.	3.0	5.2	7.4	0.444
Glass Mt., Ore.	---	---	---	---

thick walled, approximately 1½ inches thick with a 3/4-inch diameter by 9-inch-long cylindrical chamber. Two rods extended into the chamber. One rod contained a thermal couple connected to a Wheatstone bridge operating simultaneously as a thermostat for the heater and as a monitor of thermal variation which was recorded on x-y recorder. The thermal mass of the reaction vessel was approximately 35 pounds. One full day was required to bring the samples and vessel up to the equilibrated experimental temperature. At the experimental temperature, double-distilled water was let into the vessel while observing the external pressure gauge. The water was converted into the gaseous phase within the vessel up to the saturation pressure, the highest pressure at that temperature. Water was introduced after reaching this pressure such that a two-phase water/gas system was formed inside the experimental chamber. A two-phase water system simulates the natural conditions of hydration.

In the first and successive experiments, samples of each obsidian source were freshly chipped and placed in laboratory-fabricated sample capsules made by crimping 1/4-inch OD stainless steel tubing. After the first two experiments, samples were encoded with a diamond stylus for sample identification rather than using individual sample capsules.

The induced samples of both experiments were prepared for hydration measurements by Frank Findlow, of the UCLA Obsidian Hydration Dating Laboratory, under the direction of Professor Clement Meighan. No hydration bands were observed on the samples randomly prepared. The apparent discrepancy of the UCLA experiments and those of the Nasedkin experiments was discussed with many researchers; the problem was not resolved. In 1974, several of these samples were examined by the hydrogen profile technique at Cal Tech (Leich, Tombrello, and Burnett 1973; Lee et al. 1974). These samples revealed that there was a hydrated surface layer within each of the experimental samples without a birefringent layer. Further discussion of the anomaly did not elucidate any solutions. Finally, the idea that the stress birefringence was annealed during experimental cooling was suggested. With this in mind in 1974, Thomas Kaufman undertook the third experiment in the hydrothermal laboratory, under the direction of Professor G. Ernst. In this experiment, the sample chamber was degassed at the end of the experiment. Experiment 3 was a success—hydration bands were observed on most of the samples as shown in table 2-2. Perhaps the decompression of the reaction chamber and rapid cooling of the smaller reaction vessels did not anneal the birefringence of the hydration bands. The success of this experiment set the stage for more extensive series of experiments. At about the same time, Dr. I. Friedman of the United States Geological Survey provided a preprint of a paper on induced hydration, appearing as Friedman and Long (1976), which acted as a further incentive for conducting an extensive series of experiments. Thus, two nearly simultaneous events acted to rejuvenate the following hydration research program.

Using the mathematical models presented by Friedman and Long (1976) three sets of three experiments were planned. Each set of three experiments was to be conducted at a particular temperature, i.e., isothermally controlled. Each experiment of each set would be run for a particular duration. The three sets of experiments were to be conducted at the following temperatures: $150°C$, $163°C$, and $200°C$. Since experiment 3 was successful, only

Table 2-3

RESULTS OF 163°C INDUCED HYDRATION EXPERIMENTS

Source	(Exp. 3) 4 Days	(Exp. 9) 15 Days	(Exp. 7) 24 Days	$K_2 \times 10^6$ $\mu^2/10^3 \text{yr}$
Annadel	3.3	5.7	7.2	0.858
Benton Valley, W. Queen	3.2/3.5	5.6	---	0.938
Bodie Hills	3.3/3.8	5.7	8.0	1.019
Borax Lake, Area 1	---	---	---	---
Borax Lake, Area 2	---	---	---	---
Buck Mt.	2.4	5.0	---	0.567
Casa Diablo, Section 29	---	---	---	---
Coso	2.1/2.2	5.5	8.1	0.644
Cowhead Lake, Section 34	3.8/5.0	---	7.6	1.492
Cowhead Lake, 8-Mile Creek, No.	3.4/3.9	4.6	7.7	0.965
Fish Springs	3.4/4.0	7.5	7.8	1.202
Inyo Craters, Hill 8160	---	---	---	---
Inyo Craters, Hill 8491	---	---	---	---
Inyo Craters, Hill 8611N	---	---	---	---
Jawbone Canyon	---	8.1	6.1	1.081
Modoc Glass Mt., Dacite Flow	---	---	---	---
Modoc Glass Mt., Rhyolite	2.6/3.6	4.1	---	0.736
Modoc Glass Mt., Little Glass Mt.	2.9/3.3	5.5	7.7	0.850
Modoc Glass Mt., Med. L. Glass Flow	---	---	---	---
Monache Meadows	---	---	---	---
Mono Craters, North Crater	2.4	---	8.9	0.865
Mono Craters, Northwest Coulee	3.6/4.9	7.5	8.3	1.447
Mono Glass Mt., Section 17	---	---	---	---
Mt. Konocti	3.2/3.4	4.9	6.1	0.868
Napa Glass Mt., West	---	---	---	---
Napa Glass Mt., East	3.0/3.9	5.6	8.6	1.024
Napa Glass Mt., South	2.7/2.9	6.2	7.9	0.829
Napa Glass Mt., E. Dago Valley	3.1/3.8	5.5	8.3	1.081
Napa Glass Mt., W. Dago Valley	3.7/3.9	6.7	8.0	1.778
Napa Glass Mt., Hill 450*	2.7	---	7.1	0.716
Obsidian Butte, Northwest	---	---	---	---
Sugarhill, Section 21	---	---	---	---
Duck Flat, Nev.	2.4/3.6	6.9	---	0.956
Fletcher, Nev.	---	---	---	---
Mt. Hicks, Nev.	3.1/4.7	5.5	7.4	0.815
Long Valley, Nev.	---	6.5	8.5	1.063
Pine Grove Hills, Nev.	3.2/3.2	5.6/5.9	7.5/8.5	0.850/0.960
Beatty's Butte, Ore.	---	---	---	---
Glass Butte, Ore.	---	4.9	7.3	0.697
Glass Mt., Ore.	---	---	---	---

Table 2-4

RESULTS OF 172°C INDUCED HYDRATION EXPERIMENTS

Source	(Exp. 15) 3 Days	(Exp. 160) 13 3/4 Days	(Exp. 17) 28 Days	$K_3 \times 10^6$ $\mu^2/10^3 yr$
Annadel	4.7	7.0	12.7	2.030
Benton Valley, W. Queen	5.4	7.7	13.6	2.511
Bodie Hills	---	7.4	11.8	1.634
Borax Lake, Area 1	---	9.5	---	---
Borax Lake, Area 2	3.6	7.7	---	1.575
Buck Mt.	5.0	6.4	10.2	1.824
Casa Diablo, Section 29	4.8	7.7	12.8	2.172
Coso	6.0	8.4	---	3.127
Cowhead Lake, Section 34	6.0	6.9	8.6	2.203
Cowhead Lake, 8-Mile Creek, No.	4.8	7.5	---	2.148
Fish Springs	6.3	---	---	---
Inyo Craters, Hill 8160	---	---	---	---
Inyo Craters, Hill 8491	---	---	---	---
Inyo Craters, Hill 8611N	---	---	---	---
Jawbone Canyon	---	---	---	---
Modoc Glass Mt., Dacite Flow	4.2	4.9	10.9	1.444
Modoc Glass Mt., Rhyolite	---	---	---	---
Modoc Glass Mt., Little Glass Mt.	---	---	---	---
Modoc Glass Mt., Med. L. Glass Flow	---	7.1	---	---
Monache Meadows	---	8.1	---	---
Mono Craters, North Crater	9.0	---	---	---
Mono Craters, Northwest Coulee	6.5	---	---	---
Mono Glass Mt., Section 17	---	---	12.3	---
Mt. Konocti	4.4	7.0	11.4	1.782
Napa Glass Mt., West	5.3	6.4	---	2.252
Napa Glass Mt., East	5.8	9.3	---	3.194
Napa Glass Mt., South	5.7	7.1	11.4	2.328
Napa Glass Mt., E. Dago Valley	5.3	9.0	---	2.783
Napa Glass Mt., W. Dago Valley	6.0	7.3	---	2.897
Napa Glass Mt., Hill 450+	4.9	7.5	10.9	1.988
Obsidian Butte, Northwest	6.6	---	---	---
Sugarhill, Section 21	5.3	9.0	10.7	2.353
Duck Flat, Nev.	5.7	9.3	14.6	3.009
Fletcher, Nev.	---	12.9	---	---
Mt. Hicks, Nev.	5.0	10.0	14.8	2.850
Long Valley, Nev.	6.2	9.1	---	3.437
Pine Grove Hills, Nev.	---	---	---	---
Beatty's Butte, Ore.	4.2	---	14.0	2.351
Glass Butte, Ore.	4.7	7.3	3.8	2.195
Glass Mt., Ore.	5.5	7.1	12.0	1.481

Table 2-5

RESULTS OF 200°C INDUCED HYDRATION EXPERIMENTS

Source	(Exp. 4) 4 Days	(Exp. 5) 11 Days	(Exp. 6) 18 Days	$K_4 \times 10^6$ $\mu^2/10^3 yr$
Annadel	11.6	---	---	---
Benton Valley, W. Queen	---	17.7	---	---
Bodie Hills	---	---	---	---
Borax Lake, Area 1	38.0	8.8	---	---
Borax Lake, Area 2	x	x	x	---
Buck Mt.	---	17.3	---	---
Casa Diablo, Section 29	---	---	---	---
Coso	---	35.0	---	---
Cowhead Lake, Section 34	---	---	---	---
Cowhead Lake, 8-Mile Creek, No.	---	---	---	---
Fish Springs	---	---	---	---
Inyo Craters, Hill 8160	---	---	---	---
Inyo Craters, Hill 8491	---	---	---	---
Inyo Craters, Hill 8611N	---	---	---	---
Jawbone Canyon	---	---	---	---
Modoc Glass Mt., Dacite Flow	---	---	---	---
Modoc Glass Mt., Rhyolite	7.8	14.1	---	6.074
Modoc Glass Mt., Little Glass Mt.	---	19.4	---	---
Modoc Glass Mt., Med. L. Glass Flow	---	---	---	---
Monache Meadows	---	---	---	---
Mono Craters, North Crater	13.9	20.1	---	15.52
Mono Craters, Northwest Coulee	---	7.1	---	---
Mono Glass Mt., Section 17	---	---	---	---
Mt. Konocti	8.2	14.3	---	6.461
Napa Glass Mt., West	---	---	---	---
Napa Glass Mt., East	---	---	---	---
Napa Glass Mt., South	---	---	---	---
Napa Glass Mt., E. Dago Valley	---	25.0	---	---
Napa Glass Mt., W. Dago Valley	---	---	---	---
Napa Glass Mt., Hill 450+	---	---	---	---
Obsidian Butte, Northwest	---	---	---	---
Sugarhill, Section 21	---	---	---	---
Duck Flat, Nev.	12.6	---	---	---
Fletcher, Nev.	---	---	---	---
Mt. Hicks, Nev.	22.4	30.0	---	37.82
Long Valley, Nev.	9.6	19.9	---	10.77
Pine Grove Hills, Nev.	---	---	---	---
Beatty's Butte, Ore.	6.1	15.5	---	5.684
Glass Butte, Ore.	---	17.1	---	---
Glass Mt., Ore.	---	---	---	---

eight new experiments were set up. Two standardized ovens set at 150°C and 200°C were made available by the laboratory staff under Professor I. Kaplan, UCLA Department of Geochemistry. The two experiments to be run at 163°C were run on a programmable oven made available by the same laboratory. In less than seven weeks, all experiments were conducted by staggering the experimental vessels in different ovens. In these experiments, simple reaction vessels were employed which were made of 3/4-inch Swagelok pipe fittings capping 3/4-inch ID stainless stell tubing. Doubly distilled water was used which filled approximately 75% of volume of the reaction vessel. This percent of the total volume was left empty to prevent destruction of the vessel.

The results of these experiments are presented in tables 2-2, 2-3, and 2-5. The results of the 200°C experiments illustrate the anomaly of earlier experiments. Thus, a new set of experiments was set up for 172°C. These results are presented in table 2-4. In the section that follows, the mathematical models and calculation of results will be presented.

CALCULATION OF INDUCED HYDRATION RATES

Friedman and Long (1976) have illustrated some of the mathematical steps involved in evaluating the results of induced hydration experiments. A more thorough presentation of mathematical and physical steps is presented here.

The Arrhenius equation related the hydration rates to temperature in equation 2-1:

$$k = A \, \text{Exp} \, (-E/RT) \qquad [2\text{-}1]$$

where:

k = hydration rate (micrometers squared per thousand years)
A = the diffusion constant
E = activation energy (calories per mole)
R = gas constant (calories per degree per mole)
T = absolute temperature (degrees Kelvin or K)

With that predicted relationship, the experimenter can relate a series of hydration rates to their respective experimental temperatures. The determination of a hydration rate for a single experimental temperature is simply the relationship of the progression of the hydration (micrometers squared) as a function of increasing experimental time, as shown in equation 2-2:

$$\text{(a)} \quad k_T = x^2/t \qquad [2\text{-}2]$$

where:

k_T = hydration rate at temperature T (micrometers squared per thousand years)
x = hydration band thickness (microns)
t = experimental time (thousand years)

For example, if three experiments were performed at a particular temperature over three different periods of time, the three observed hydration band thicknesses could be related to the experimental times by using equation 2-2. By simple linear regression, the hydration rate at this experimental temperature is the slope of the relationship between the observed hydration thickness squared and the experimental times.

The Arrhenius equation requires the determination of several hydration rates for several experimental temperatures. Taking the natural logarithm, the Arrhenius equation has the following form:

$$\ln k = \frac{-E}{RT} + \ln A \qquad [2\text{-}3]$$

The calculation of the source-specific induced activation energies and the diffusion constants were calculated using equations 2-2 and 2-3. First, the experimental hydration rates, "k_T," at temperature T were calculated for all sources. The results of these calculations are tabulated in tables 2-2 through 2-5. Secondly, these values and the corresponding temperatures of each experiment were entered into equation 2-3. The source-specific activation energies, E, and diffusion constants, A, were calculated and are tabulated in table 2-6.

The range of variation of results of the induced hydration experiments again validates the necessity of source specification to further refine the obsidian hydration dating technique. This finding supports the major thesis of the dissertation that the hydration rates vary among obsidian sources. Also, the success of the induced hydrothermal experiments indicate that this technique may be quite useful for regional hydration rate estimation.

It is now important to examine the degree of concordance between empirical and induced hydration rates.

The experimental technique of inducing the hydration of obsidian promises to be quite useful. However, we are not sure whether the accelerated hydration process activated at elevated temperatures and pressures is equivalent to the "natural" hydration process observed in obsidian artifacts. In this section, the degree of concordance or correlation between the two results is examined. If both processes are identical, a high correlation will be shown. If the processes are similar, a high correlation with nonrandom error will be indicated. If the processes are quite different, the converse will be observed.

Thus, the induced hydration rate constants were calculated using equation 2-1 in conjunction with the factors in table 2-6.

CONCLUSIONS

In conclusion, the results of the induced hydration indicate a great deal of variability in the source-specific activation energies and diffusion coefficients, which was expected. This variability is in part an artifact of the sample size. A larger number of experiments over ambient times and temperatures would have been preferable. Nevertheless, this technique is useful at least to establish the rank-order of the hydration rates of a series of obsidian sources.

Table 2-6

COEFFICIENTS FOR SOURCE-SPECIFIC INDUCED HYDRATION EQUATIONS

Source	$150°C$ $K_1 \times 10^6$	$163°C$ $K_2 \times 10^6$	$172°C$ $K_3 \times 10^6$	$200°C$ $K_4 \times 10^6$	A $\mu^2/10^3 yr$	E Kcal/mole
Annadel	0.630	0.858	2.030	---	3.62×10^{15}	18.97
Benton Valley, W. Queen	0.311	0.938	2.511	---	4.30×10^{23}	35.14
Bodie Hills	---	1.019	1.634	---	1.40×10^{16}	20.22
Borax Lake, Area 1	0.625	---	---	---	---	---
Borax Lake, Area 2	---	---	1.575	---	8.19×10^{13}	15.71
Buck Mt.	0.370	0.567	1.824	---	7.60×10^{18}	25.88
Casa Diablo, Section 29	---	---	2.172	---	---	---
Coso	0.346	0.644	3.127	---	8.95×10^{23}	35.79
Cowhead Lake, Section 34	0.398	1.492	2.203	---	1.01×10^{21}	29.34
Cowhead Lake, 8-Mile Creek, No.	0.334	0.965	2.148	---	6.15×10^{21}	31.49
Fish Springs	0.326	1.202	---	---	3.28×10^{24}	36.78
Inyo Craters, Hill 8160	---	---	---	---	---	---
Inyo Craters, Hill 8491	---	---	---	---	---	---
Inyo Craters, Hill 8611N	---	---	---	---	---	---
Jawbone Canyon	0.322	1.081	1.444	---	1.40×10^{23}	34.13
Modoc Glass Mt., Dacite Flow	---	---	---	6.074	---	---
Modoc Glass Mt., Rhyolite	0.218	0.736	---	---	5.63×10^{18}	25.85
Modoc Glass Mt., Little Glass Mt.	0.319	0.850	---	---	5.97×10^{19}	27.62
Modoc Glass Mt., Med. L. Glass Flow	---	---	---	---	---	---
Monache Meadows	---	---	---	15.52	---	---
Mono Craters, North Crater	0.541	0.865	---	---	9.23×10^{19}	27.73
Mono Craters, Northwest Coulee	0.364	1.447	---	---	4.56×10^{25}	38.90
Mono Glass Mt., Section 17	---	---	---	---	---	---
Mt. Konocti	0.273	0.868	1.782	6.461	2.02×10^{18}	24.75

(contd.)

Table 2-6 (contd.)

Source	150°C $K_1 \times 10^6$	163°C $K_2 \times 10^6$	172°C $K_3 \times 10^6$	200°C $K_4 \times 10^6$	A $\mu^2/10^3 \text{yr}$	E Kcal/mole
Napa Glass Mt., West	---	---	2.252	---	---	33.55
Napa Glass Mt., East	0.424	1.024	3.194	---	8.39×10^{22}	32.83
Napa Glass Mt., South	0.326	0.829	2.328	---	2.82×10^{22}	32.75
Napa Glass Mt., E. Dago Valley	0.396	1.081	2.783	---	3.17×10^{22}	31.89
Napa Glass Mt., W. Dago Valley	0.435	1.778	2.897	---	1.26×10^{22}	27.81
Napa Glass Mt., Hill 450+	0.369	0.716	1.988	---	7.90×10^{19}	---
Obsidian Butte, Northwest	---	---	---	---	---	---
Sugarhill, Section 21	---	---	2.353	---	---	37.00
Duck Flat, Nev.	0.329	0.956	3.009	---	4.04×10^{24}	---
Fletcher, Nev.	---	---	---	---	---	36.29
Mt. Hicks, Nev.	0.476	0.815	2.850	37.82	1.94×10^{24}	29.83
Long Valley, Nev.	0.559	1.063	3.437	10.77	1.28×10^{21}	15.64
Pine Grove Hills, Nev. A	0.488	0.850	---	---	5.88×10^{13}	27.64
Pine Grove Hills, Nev. B	0.360	0.960	---	---	6.92×10^{19}	16.50
Beatty's Butte, Ore.	0.707	---	2.351	5.684	2.56×10^{14}	25.98
Glass Butte, Ore.	0.444	0.697	2.195	---	1.03×10^{19}	---
Glass Mt., Ore.	---	---	1.481	---	---	---

Chapter 3

INTRINSIC VARIABLES OF THE HYDRATION PROCESS

INTRODUCTION

The hypothesis presented by Ericson and Berger (1976) predicts that the intrinsic properties of individual obsidian sources will account for the observed variations in the hydration rates among obsidian sources. This hypothesis was derived from observations of several associated groups of obsidian artifacts under identical hydration conditions—time, temperature, and soil conditions. These groups showed significant differences in hydration when regrouped by their specific source of origin. However, apart from this generalized observation the specific effects of the intrinsic variables on the hydration process remain unknown and unidentified.

The isolation of the major intrinsic variables of the hydration process is important for a greater understanding of the hydration process, a very complex phenomenon, and its variation. Perhaps as important, such information would permit the archaeologist on a worldwide basis to compare chronometric rates with calculated rates, derived as functions of intrinsic and environmental variables.

In order to isolate the major intrinsic variables of the hydration process, an extensive research project was undertaken over the last five years with the source-specific hydration rates previously described in chapter 1. The rates were considered as the dependent variable and the intrinsic properties of each obsidian source as independent variables in multiple regression analysis. For these purposes, obsidian source samples were subjected to a series of physical measurements, initially summarized by Ericson et al. (1975).

The basis of the selection of the intrinsic variables was governed by two research concerns. First, there was a concern to present the most extensive physical description of the source samples possible. This is a normal procedure, which is used for isolating the effects of independent variables on a particular phenomenon. Second, there was a concern to present a series of variables based on theoretical considerations of the hydration phenomena. As a result, there are two types of intrinsic variables. The physical properties which were directly measured have been termed <u>empirical intrinsic variables</u>. The second type of variables, which were derived from the measurements of the physical properties, have been termed <u>calculated intrinsic variables</u>. The measurement or calculation of each intrinsic variable required a specific set of experimental procedures or mathematical calculations. These procedures and details of variable selection are presented in the following sections.

Table 3-1

MAJOR OXIDE ANALYSIS OF OBSIDIAN SAMPLES (Weight-percent)

Source	Sample/Area	Sample #	SiO$_2$	Al$_2$O$_3$	CaO	Na$_2$O	K$_2$O	Fe$_2$O$_3$	MgO	TiO$_2$	Total
Obsidian Butte	Southwest	1-1-1-1	75.101	12.178	0.746	4.242	4.296	2.600	0.131	0.154	99.348
	Southeast	1-1-2-5	74.605	12.252	0.802	4.007	4.235	2.675	0.148	0.142	98.866
	Northeast	1-1-3-1	72.624	12.580	0.808	3.887	4.173	2.615	0.113	0.163	96.963
	Northwest	1-1-4-1	73.455	11.513	0.472	3.745	4.385	2.210	0.048	0.073	95.901
Coso		2-1-1-17	78.367	12.404	0.491	3.571	4.454	1.236	0.089	0.053	100.665
Fish Springs		3-10-11	78.402	12.883	0.510	3.808	4.686	0.011	0.042	0.036	101.178
Inyo Craters	Hill 8491	4-1-0-6	69.511	14.328	1.069	3.707	4.841	2.289	0.235	0.225	96.205
	Hill 8160	4-2-0-3	68.274	13.582	1.009	3.594	5.159	2.149	0.180	0.181	94.128
	Hill 8520	4-3-0-2	62.373	13.764	1.686	3.790	4.603	2.928	0.325	0.398	89.867
	Hill 8611S	4-5-1-2	74.478	13.302	0.725	3.794	5.448	1.811	0.106	0.095	99.759
	Hill 8611N	4-5-2-1	74.501	13.523	0.720	3.872	5.480	1.924	0.098	0.103	100.221
Mono Craters	Northwest Coulee	4-1-0-7	79.376	12.142	0.602	3.300	4.718	1.288	0.034	0.058	101.518
	North Crater	5-2-0-1	77.251	12.417	0.483	3.644	4.638	1.305	0.044	0.033	99.815
Mono Glass Mt.	Section 24	6-1-0-3	79.227	12.240	0.512	3.313	4.658	0.012	0.083	0.066	99.911
	Section 17	6-1-1-6	77.609	11.948	0.377	3.325	4.908	0.865	0.037	0.037	99.106
West Queen	Truman Canyon	7-1-0-10	79.775	12.983	0.643	3.749	4.812	0.804	0.124	0.101	103.081
	West Queen	7-2-0-6	70.293	12.404	0.464	3.476	4.552	0.923	0.057	0.062	92.231
Duck Flat, Nev.		8-1-0-5	82.260	11.628	0.312	3.834	4.694	3.141	0.048	0.111	105.028
Buck Mt.		9-1-1-1	81.950	13.326	0.889	3.577	4.956	0.858	0.126	0.107	105.781
		9-1-1-5	76.098	12.749	0.909	3.219	4.585	0.628	0.149	0.104	98.641
		9-1-1-13	78.384	13.058	0.910	3.388	4.626	0.830	0.117	0.103	101.416
		9-1-1-29	78.406	12.079	0.871	3.200	4.663	0.851	0.082	0.106	101.258
	Bucher Creek	9-1-3-3	80.867	13.287	0.900	3.537	4.816	1.015	0.136	0.130	104.688
Cowhead	Section 32	10-1-1-1	76.332	12.171	0.318	3.704	4.052	0.877	0.048	0.030	97.532
	Section 33	10-1-2-3	77.486	12.376	0.296	4.035	3.856	0.859	0.048	0.028	98.984

(contd.)

Table 3-1 (contd.)

Source	Sample/Area	Sample #	SiO$_2$	Al$_2$O$_3$	CaO	Na$_2$O	K$_2$O	Fe$_2$O$_3$	MgO	TiO$_2$	Total
Cowhead continued	Section 34	10-1-3-4	77.106	12.184	0.315	4.086	4.043	0.869	0.073	0.030	98.706
	8-Mile Creek S	10-2-1-1	78.105	12.353	0.317	4.023	4.170	0.859	0.051	0.030	100.218
	8-Mile Creek N	10-2-2-1	79.659	12.607	0.325	3.855	4.241	0.903	0.037	0.030	101.657
Sugarhill	Section 15	11-1-1-1	79.123	12.876	0.784	3.295	4.959	1.206	0.155	0.124	102.612
	Section 21	11-1-2-5	77.754	12.640	0.751	3.124	4.979	1.264	0.127	0.138	100.777
	Section 21	11-1-2-11	78.862	12.954	0.625	3.146	4.945	1.074	0.067	0.090	101.763
	Section 28	11-1-3-3	81.360	12.449	0.617	3.643	5.145	1.118	0.111	0.105	104.548
	Section 28	11-1-3-7	79.794	13.143	0.940	3.245	4.924	1.203	0.163	0.167	103.579
Modoc Glass Mt.	Dacite Flow	12-1-0-1	72.750	13.369	1.740	3.633	4.187	2.132	0.286	0.271	98.368
	Rhyolitic Flow	12-2-0-9	73.754	13.466	1.415	3.387	4.278	2.081	0.221	0.263	98.855
Medicine Lake	Glass Flow	12-3-0-1	64.703	14.402	3.938	3.219	3.284	3.631	0.839	0.421	94.337
Little Glass Mt.		12-4-0-7	65.530	13.301	1.101	3.654	3.918	3.031	0.149	0.183	89.867
Mt. Konocti		13-1-0-5	78.441	13.436	1.332	3.311	4.690	1.699	0.262	0.257	103.428
	Sugarloaf	13-2-0-1	74.724	13.144	1.287	3.038	4.691	1.603	0.218	0.244	99.949
	McIntre Creek	13-3-0-3	75.045	13.041	1.148	3.135	4.718	1.607	0.224	0.232	99.210
Borax Lake	Area 1	13-4-1-3	81.729	12.044	0.730	3.188	4.877	1.324	0.136	0.064	104.092
	Area 2	13-4-2-1	80.732	12.855	0.844	3.277	4.708	1.265	0.193	0.092	104.006
Napa Glass Mt.	West	14-1-1-1	74.486	12.063	0.422	3.572	4.568	1.385	0.108	0.085	96.690
	East	14-1-2-1	79.255	12.409	0.414	3.959	3.373	1.472	0.030	0.087	101.999
	South	14-1-3-2	66.000	14.556	0.265	4.045	3.834	1.323	0.015	0.052	90.090
	E. Dago Valley	14-2-0-4	79.172	12.968	0.310	4.163	4.492	1.515	0.033	0.063	102.716
	W. Dago Valley	14-2-0-1	79.468	12.442	0.345	4.163	4.475	1.492	0.047	0.065	102.497
	Hill 450+	14-4-0-3	76.833	12.047	0.329	3.914	4.436	1.505	0.022	0.064	99.150
Annadel		15-1-0-1	70.056	12.021	0.995	2.775	4.265	2.072	0.093	0.165	93.443
Monache		14-1-0-6	76.704	13.637	1.136	4.550	3.664	2.640	0.112	0.144	107.407
Monache Meadows		16-1-0-1	77.420	14.076	0.723	3.823	4.400	0.700	0.073	0.034	101.249

(contd.)

Table 3-1 (contd.)

Source	Sample/Area	Sample #	SiO_2	Al_2O_3	CaO	Na_2O	K_2O	Fe_2O_3	MgO	TiO_2	Total
Beatty's Butte, Ore.		17-1-0-4	63.803	14.581	0.934	3.485	4.294	1.392	0.125	0.119	88.733
Long Valley, Nev.		18-1-0-1	75.615	10.704	0.242	3.750	4.748	2.301	0.00	0.087	97.497
Glass Mt., Ore.		19-1-0-6	70.587	12.791	0.454	4.027	4.578	7.700	0.044	0.180	95.361
		19-2-2-2	73.819	12.054	0.466	4.159	4.761	2.683	0.062	0.184	98.188
Glass Butte, Ore.		20-1-0-1	79.688	12.513	0.493	3.495	4.253	0.906	0.083	0.059	101.490
Rustler Canyon		21-1-0-0	76.816	11.778	0.200	3.904	4.980	1.445	0.063	0.134	99.320
Shoshone	Dublin Hills	22-1-0-12	69.627	11.870	0.766	2.913	4.853	1.251	0.164	0.161	91.615
	Jubilee Pass	22-3-0-5	72.764	12.058	1.438	2.839	3.871	1.311	0.270	0.136	94.687
	Charley Brown (vitrophyre)	22-4-0-11	77.642	12.716	0.883	3.023	4.858	1.019	0.313	0.073	100.527
	Charley Brown (obsidian)	22-4-0-16	70.394	13.823	0.581	3.406	4.219	1.048	0.158	0.065	93.694
Jawbone Canyon		23-1-0-2	63.666	12.930	0.507	2.650	4.285	0.004	0.074	0.033	84.009
Casa Diablo	Section 35	24-1-1-1	77.558	12.752	0.795	2.976	5.139	1.500	0.104	0.165	95.989
	Section 29	24-1-2-1	75.764	12.806	0.702	3.338	5.193	1.458	0.038	0.115	99.414
	Section 22	24-1-3-11	74.775	12.624	0.760	3.084	5.250	1.529	0.115	0.134	98.271
Mt. Hicks, Nev.		25-1-1-1/6	83.401	12.728	0.481	3.242	4.897	0.717	0.059	0.058	105.583
Fletcher, Nev.		26-1-1-1/8	75.901	12.290	0.655	3.389	4.726	0.794	0.051	0.073	97.879
Bodie Hills		29-1-1-3	73.163	12.006	0.627	3.152	4.642	0.775	0.086	0.068	94.519
Pine Grove Hills, Nev.		28-1-1-3	77.269	11.864	0.749	3.030	4.814	0.733	0.052	0.058	98.569

Table 3-2

X-RAY FLUORESCENCE SPECTROGRAPHIC CONDITIONS

Conditions	Si	Al	Ca	Fe	Na	K	Mg	Ti
Primary Rad.	Cr	Cr	Cr	Cr	Cr	Cr	Cr	Cr
Voltage (KV)	50	50	50	50	50	50	50	50
Amperage (ma)	25	25	25	20	28	25	28	25
Crystal	PET	PET	PET	LiF(200)	KAP	PET	KAP	PET
Analytical Line	$K\alpha$	$K\alpha$	$K\alpha$	$K\alpha$	$K\alpha$	$K\alpha$	$K\alpha$	$K\alpha$
Window	Al-Mylar	Al-Mylar	Al-Mylar	Al-Mylar	Polypropy	Al-Mylar	Polypropy	Al-Mylar
Peak (2)	79.10	115.05	57.10	57.05	23.60	21.55	13.70	6.79
Background (20)	83.00	106.00	19.00	60.00	24.50	19.00	17.00	9.5
Counting Period (seconds)	10	20	10	10	100	10	50	10
Number of Counting Periods	3	3	3	3	3	3	3	3
Detector	Scint.	Scint.	Scint.	Scint.	Scint.	Scint.	Scint.	Scint.
Standard (G-2) Wt. % Oxide	69.20	15.42	1.98	2.67	4.05	4.46	0.76	0.47
Precision (%) of Measurement	1.51	1.24	0.88	1.34	3.00	1.64	1.61	1.76

EMPIRICAL INTRINSIC VARIABLES

The physical properties of a sample of each obsidian source were measured to present an extensive physical description of the obsidian. In addition, these measurements provided a means to determine the degree of variability of the properties among the sources under examination. In all, 15 physical properties were examined. They are, in their order of presentation: (1-8) weight-percentage of the eight major oxides, (9) density, (10) hardness, (11) weight-percent of the water in the hydrated samples, (12) internal water concentration, weight percent, (13) initial structural bonding of water, (14) final structural bonding of water, and (15) degree of crystallization.

Variables 1-8: Weight Percent of the Eight Major Oxides

One of the most important properties of any material is its chemical composition. In general, changes in the chemical composition preclude changes in most of the other physical properties, if not the changes in the classification of the material itself. For these reasons, the chemical composition of the obsidian samples were determined by X-ray fluorescence analysis. The eight major oxides which were analyzed are in their order of presentation as shown in table 3-1: (1) silicon oxide, SiO_2; (2) aluminum oxide, Al_2O_3; (3) calcium oxide, CaO; (4) sodium oxide, Na_2O; (5) potassium oxide, K_2O; (6) iron oxide, presented as Fe_2O_3 only; (7) magnesium oxide, MgO; (8) titanium oxide, TiO_2; to this list the total wt % has been added.

The obsidian source samples were prepared for analysis by cutting with a diamond-charged brass saw blade, grinding in a 6-inch diameter tungsten carbide mortar mounted in a rotary Spex shatterbox, passing the resulting powder through a # 275 mesh copper sieve, weighing 1.0000 gram amounts, adding 0.2000 gram USP grade methyl cellulose as a powder bonding agent, mixing the obsidian powder and methyl cellulose mixture for five minutes in a tungsten carbide Spex mixer mill # 8000, and compressing this mixture at 22 tons/in^2 inch 1 1/4 -inch diameter pellets formed by a surrounding methyl cellulose backing. (It should be noted that careful standardization of the analytical samples is a prerequisite to the accuracy of the X-ray fluorescence analysis.) The three major contributions to analytical error are: (1) the quality of the prepared surface of the sample, (2) the homogeneity of the sample itself, and (3) the chemical and physical similarity of the standard used for comparison and the analyzed material.

The obsidian samples were analyzed using a Norelco X-ray fluorescence spectrometer. The actual instrumental parameters for each element are presented in table 3-2. In a normal experiment, seven samples and the standard, USGS G-2, were placed in the sample chamber. The chamber was then evacuated to a pressure of 5-10 microns Hg. The characteristic $K\alpha$ intensities of each element were counted for fixed time intervals, varying from 10 to 100 seconds, depending on the accumulation of gross counts. All samples within the chamber were counted in succession, leaving the goniometer dial in a fixed 2θ position. Likewise, the background intensity for each element was counted with the goniometer fixed at the predetermined 2θ value. The adoption of this stepwise procedure insured the most accurate comparison between a set of analytical examples and the standard. This modification of the procedure provided a means to minimize the delay time between the analysis of any given sample and the standard, thus minimizing error due to electronic fluctuations of the system. In addition, the measurement of the intensity at a fixed 2θ position allowed for the exact reduplication of the position of the counter in 2θ space. Following the accumulation of intensity data, the weight percent of each element of each sample was calculated by taking the mean of the elementary Kα intensity, subtracting the associated background intensity, and converting the corrected intensity to weight percent oxide by using the corrected intensity of the standard USGS-G-2, the chemical composition of which is described by Flanagan (1973).

The results of the analysis of the 69 obsidian samples are shown in table 3-1. The criteria for evaluating the results of the analysis of a given sample is that the total weight percent should sum to approximately 100 %. A total in the range from 97 to 101 wt % is considered to be a reliable analysis. This range is explained by considering two contributing factors: (1) the error due to total precision of measurement is estimated at ± 1.50, shown in table 3-2, and (2) the total weight percent of eight major oxides exclude contributions from water, minor elements, and trace elements which are estimated to comprise an additional 1 to 3 weight percent to the total.

In summary, the major elementary composition is considered to be an important variable set influencing the hydration process. The X-ray fluorescence analyses of 69 obsidian source samples are presented as variables (1-8) in table 3-1.

Variable 9: Density

The density is a composite physical property of two independent variables, namely, volume and mass. This relationship is formulated in equation 3-1:

$$p = \frac{m}{v} \qquad [3-1]$$

where:

p = density (grams /cc)
m = mass (grams)
v = volume (cc)

Although mass as an independent variable is determined by the quantity of chemical constituents, the volume of the material is dependent on the structure, i.e., on how the chemical constituents are averaged or packed within the structure. There are two qualitative states of the density: the material has either an annealed or unannealed density. The annealed density is the maximum density attained by a material for a particular state of entropy and environmental conditions. By definition, the unannealed density is a value less than the annealed value. The inverse relationship between density and volume predicts that a lower, unannealed density indicates a larger expanded volume of the material.

It is expected that the relative expansion of the volume indicates both a greater space for "water" diffusion and an expanded structure which would facilitate bonding of the migrating "water" species. In order to normalize the density of a given sample, whether annealed or unannealed, a calculated variable has been determined in a later section of this chapter—the Ω factor.

Since glasses are high entropy materials, the structures are neither as highly ordered nor as dense as their crystalline counterparts. Obsidian, a natural glass, is formed by volcanic processes. Its densification is a function of the processes of its formation. The density components of obsidian, suggested by Ericson et al. (1975:136), are revised here as equation 3-2:

$$p = ap_g + bp_x + cp_p + dp_v + ep_n \qquad [3\text{-}2]$$

where:

p = the density
ap_g = the density component of the glass phase
bp_x = the density component of the crystals
cp_p = the density component of the pore structure
dp_v = the density component of the volatiles
ep_n = the density component of quality of annealing

The densities of the obsidian samples, reported in table 3-3, were measured by cutting small sample plates, measuring the weight in distilled water and air on a Troemner specific gravity chain balance (Model S-100) and calculating the density using the formula presented in equation 3-3:

$$p = \frac{w_1 - w_2}{(w_1 - w_2)(w_3 - w_4)} \qquad [3\text{-}3]$$

where:

p = the density (specific gravity)
w_1 = weight of sample + system suspended in air
w_2 = weight of system suspended in air
w_3 = weight of sample + system suspended in water
w_4 = weight of system suspended in water

The density measurements were recorded five times to establish the mean and standard deviation of the observations.

Variable 10: Hardness

The hardness of a material measures the pressure necessary to break or elastically deform the structure. The hardness provides a measure of the average nature of the chemical bonds which form the structure of the material. It has been suggested that the hardness in obsidians is a complex function of the crystalline phases, pores, and chemical composition (Ericson et al. 1975:139).

There are a number of theoretical considerations which suggest that the hardness, as a measure of the relative rigidity of the structure of obsidian, may be an important variable of the hydration process. The ability of the structure to withstand the strain produced by the addition of water in the hydration process (Friedman and Smith 1960) would retard the formation and

Table 3-3

PHYSICAL PROPERTIES OF OBSIDIAN SAMPLES

Source	Sample/Area	Sample Number	Density (g/cc)	Hardness (Kg/mm^2)	Degree of Percent Crystallization			Sample Concentration of Water (Wt %)			
						No.	Max.	Internal	No.	Max.	Internal
Obsidian Butte	Southwest	1-1-1-1	---	---	4	+4	3.190	0.6648	-5	2.725	1.0616
	Southeast	1-1-2-5	---	562 ± 85	2	-3	2.975	0.7077	-4	0.7595*	0.7104
	Northeast	1-1-3-1	---	---	25	-2	0.3663*	0.8149	-3	0.411*	0.7712
	Northwest	1-1-4-3	---	---	7	-1	0.6166*	0.8203	-4	1.7515	1.2680
Coso		2-1-1-17	2.3576	773 ± 38	10	-4	3.735	1.2171	25	8.165	0.7401
Fish Springs		3-10-11	2.3642	714 ± 39	2	-5	4.0928	0.5737	---	---	---
Inyo Craters	Hill 8491	4-1-0-6	2.3112	---	90	-8	2.475	0.8203	---	---	---
	Hill 8160	4-2-0-3	2.4372	---	100	-1	2.654	1.1965	-2	3.583	1.3047
	Hill 8520	4-3-0-2	2.4619	---	50	-1	2.785	0.5987	---	---	---
	Hill 8611S	4-5-1-2	2.3247	---		-1	4.567	1.308	---	---	---
	Hill 8611N	4-5-2-1	2.3439	---	80	-4	2.895	0.6577	---	---	---
Mono Craters	Northwest Coulee	5-1-0-7	2.3433	---	30	-2	2.251	0.5492	-4	3.163	0.6612
	North Crater	5-2-0-1	2.3628	756 ± 75	5	-3	---	1.5549	-5	5.751	0.7388
Mono Glass Mt.	Section 14	6-1-0-3	2.4265	750 ± 38	10	-1	4.5128	0.5790	-4	3.2081	0.5755
	Section 17	6-1-1-4	---	789 ± 41	5	-2	3.1813	0.6639	---	---	---
West Queen	Truman Canyon	7-1-0-16	2.3571	746 ± 69	1-4	-3	4.6916	0.7220	---	---	---
	West Queen	7-2-0-6	2.3412	---	2	-1	2.5021	0.6684	---	---	---
Duck Flat, Nev.		8-1-0-5	2.3700	832 ± 15		-9	9.6066**	0.5084	-11	3.2528	1.0527
Buck Mt.		9-1-1-1	2.3831	---	55	-2	4.0928	0.6121	---	---	---
		9-1-1-5	2.4297	697 ± 40	3	-3	2.7434	0.4959	---	---	---
		9-1-1-13	2.3633	712 ± 103		-x	4.1643	0.5960	---	---	---
		9-1-1-29	2.4014	---	15	-20	3.4494	0.5701	-26	2.8775	0.3797

(contd.)

* Reject, maximum concentration values not hydrated.
** Reject, anomalous.

Table 3-3 (contd.)

Source	Sample/Area	Sample Number	Density (g/cc)	Hardness (Kg/mm^2)	Degree of Percent Crystalli- zation	Sample Concentration of Water (Wt %)					
						No.	Max.	Internal	No.	Max.	Internal
Cowhead Lake	Bucher Creek	9-1-3-3	2.3680	---		---	---	---	---	---	---
	Section 32	10-1-1-1	2.3424	698 ± 59	10	-2	0.5272*	1.3699	-3	4.2179	0.9418
	Section 33	10-1-2-3	2.3857	---	3	-4	5.7729	0.7712	-5	3.1545	0.9874
	Section 34	10-1-3-4		---		-1	3.8694	0.6246	---	---	---
	8-Mile Creek S	10-2-1-1	2.3679	---	17	-5	3.4226	0.6907	---	---	---
	8-Mile Creek N	10-2-2-1	2.3449	---	20	---	---	---	---	---	---
Sugarhill	Section 15	11-1-1-1	2.4063		90	-2	3.3958	0.4566	---	---	---
	Section 21	11-1-2-5	2.3765		70	-4	2.8417	0.5263	---	---	---
	Section 21	11-1-2-11	2.4087	---	35	-7	4.2043	0.5495	---	---	---
	Section 28	11-1-3-3	2.3572	n.d.	2	-1	5.0311	0.7086	-4	4.0213	0.7703
	Section 28	11-1-3-7	2.4635	601 ± 39	98	---	---	---	---	---	---
Modoc Glass Mt.	Dacite Flow	12-1-0-1	2.4186	n.d.	90	-2	3.9945	0.5701	-4	3.1545	0.8391
	Rhyolitic Flow	12-2-0.9	2.4031	742 ± 34	1	-4	5.8086	1.0643	-5	3.7890	0.7810
Medicine Lake	Glass Flow	12-3-0-1	2.3882	n.d.	75	---	---	---	---	---	---
Little Glass Mt.		12-4-0-7	2.3442	625 ± 26	12	-5	2.3860	0.8918	-8	0.6612*	0.5397
Mt. Konocti		13-1-0-5	2.3560	485 ± 61	1	-3	4.5396	0.8239	-6	3.4405	1.0464
	Sugarloaf	13-2-0-1	2.3980	n.d.	20	-2	5.0133	0.6496	-3	2.6362	0.6309
	McIntre Creek	13-3-0-3	2.3752	597 ± 40	10	-1	3.3739	0.7828	-4	7.4172	1.4287
Borax Lake	Area 1	13-4-1-3	2.2288	n.d.	5	-2	2.8149	0.7998	-3	3.1634	0.7211
	Area 2	13-4-2-1	2.3567	734 ± 12	3	-3	5.4884	0.6591	---	---	---
Napa Glass Mt.	West	14-1-1-1	2.3780	835 ± 17	30	-2	4.0035	0.8855	-3	4.1375	0.7220
	East	14-1-2-1	2.3654	716 ± 73	2	-6	3.7622	0.7917	-11	0.2859*	0.5183
	South	14-1-3-2	2.4072	880 ± 12	4	-3	2.7792	0.4886	-5	6.0320	0.6058
	E. Dago Valley	14-2-0-4	2.3703	667 ± 71	5	-1	4.2269	1.5433	-2	3.1634	1.2090

(contd.)

* Reject, maximum concentration values not hydrated.

Table 3-3 (contd.)

Source	Sample/Area	Sample Number	Density (g/cc)	Hardness (kg/mm^2)	Degree of Percent Crystallization	Sample Concentration of Water (Wt %)					
						No.	Max.	Internal	No.	Max.	Internal
Napa Glass Mt. (continued)	W. Dago Valley Hill 450+	14-3-0-1	2.3770	791 ± 19	10	-1	4.1822	0.8899	-4	3.1277	1.1501
		14-4-0-3	2.3657	842 23	4	-4	6.1393	0.5746	-5	4.7720	0.6166
Annadel		15-1-0-1	n.d.	n.d.	95	-10	7.2503	0.4513	-15	6.1000	0.2158
		15-1-0-16	2.4416	590 ± 9	80	---	--	--	---	--	--
Monache Md.		16-1-0-1	--	--		-4	6.8861	3.8967	---	--	--
Beatty's Butte, Ore.		17-1-0-4	--	--		-1	2.3055	0.6630	-2	7.8049	0.9321
Long Valley, Nev.		18-1-0-1	--	--		-1	6.4682	0.6464	---	--	--
Glass Mt., Ore.		19-1-0-6	--	--		-2	4.0481	1.6148	---	--	--
		19-2-0-2	--	--		---	--	--	---	--	--
Glass Butte, Ore.		20-1-0-1	--	--		-1	3.3400	0.5191	---	--	--
Rustler Canyon		21-1-0-0	--	--	27	-3	2.4128	0.9740	-6	1.8587	0.7435
Shoshone	Dublin Hills	22-1-0-12	--	689 ± 73	95	---	--	--	---	--	--
	Jubilee Pass	22-3-0-5	--	572 + 27		-1	3.9766	5.3707	---	--	--
	Charley Brown (vitrophyre)	22-4-0-11	--	--	25	---	--	--	---	--	--
	Charley Brown (obsidian)	22-4-0-16	--	--		---	--	--	---	--	--
Jawbone Canyon		23-1-0-2	--	546 ± 32	25	-1	5.8980	7.5521	-6	1.6174	3.6192
Casa Diablo	Section 35	24-1-1-1	n.d.	799 + 67	17	-3	2.8953	0.6246	---	--	--
	Section 29	24-1-2-1	n.d.	n.d.	20	-5	2.8775	1.2877	-6	2.7077	0.6094
	Section 22	24-1-3-11	n.d.	n.d.	5	-1	3.1009	0.7166	-3	2.5468	1.2073
Mt. Hicks, Nev.		25-1-1-1/6	n.d.	874 ± 72	3	-4	3.8158	0.6836	-758	4.2717	0.3595
Fletcher, Nev.		25-1-1-1/8	n.d.	793 22	0	-3	3.9052	1.3824	-7	3.5924	1.2296
Bodie Hills		27-1-1-5	n.d.	907 8	1	-4	3.5656	0.6282	-7	2.6094	0.6738
Pine Grove Hills, Nev.		28-1-1-3	n.d.	671 76	50	-14	4.5218	0.5656	-19	4.5843	0.7962

progression of the hydration layer as a function of time. Secondly, it is not well understood why there exist saturation levels of the "water" concentration in the hydration layer (Friedman and Smith 1960:482; Lee et al. 1974:47). A possible suggestion is that the saturation levels are controlled by the internal strength of the obsidian structure, measured by its hardness value.

For these reasons, the Vickers hardness of the obsidian samples were measured by using the following procedures. The individual samples were cut into square plates, ground and polished to a mirrorlike surface, obtained with 6μ diamond paste, 0.5μ and 0.05μ Al_2O_3 powder. Each sample was indented with a Vickers hardness instrument using a 100-gram load and diamond point indenter. The diagonals of the base of the tetrahedron formed by the diamond on impact were measured by a filar micrometer at 100 magnification. One of four indentations of each three repetitive groups was selected for measurement. The Vickers hardness was calculated using the formula shown in equation 3-4:

$$H_v = \frac{2L \sin \theta/2}{d^2} \qquad [3\text{-}4]$$

where:

H_v = Vickers hardness (Kg/mm^2)
L = load (Kg)
θ = angle of crystalline faces of diamond tetrahedron (= $136°$)
d = average length of diagonal of base of tetrahedron formed by indentation (mm)

The results of these measurements are reported in table 3-3.

Variable 11: Maximum Water Concentration in the Hydration Layer

The maximum water concentration in the hydrated layer in obsidian samples is considered to be an important variable of the hydration process. In diffusion studies, the maximum concentration of the diffusing species within a medium is always considered one of the variables of the diffusion process. Thus, in the hydration process, which has been characterized as an autocatalytic phenomenon having both diffusion and reaction components (Ericson, MacKenzie, and Berger 1976; Ericson 1975), this variable might be shown to be an important intrinsic parameter.

In the simple diffusion model suggested for the hydration process by Friedman and Smith (1960) or Friedman, Smith, and Long (1966), only one water concentration plateau is predicted. However, actual measurement of the water concentration profile at depth with hydrated obsidian samples shows the presence of two concentration plateaus. For example, the concentration of the first plateau is 1.5-2.4 wt % water, and 2.1-3.7 wt % water, for the second plateau as shown in plate 3-1. The agreement between optical and concentration measurements of the hydration layer (Lee et al. 1974) suggests that the first concentration plateau is sufficient to develop the internal strain producing the stress birefringence, optically measured as the hydration layer. The influence of the second concentration plateau on the hydration process

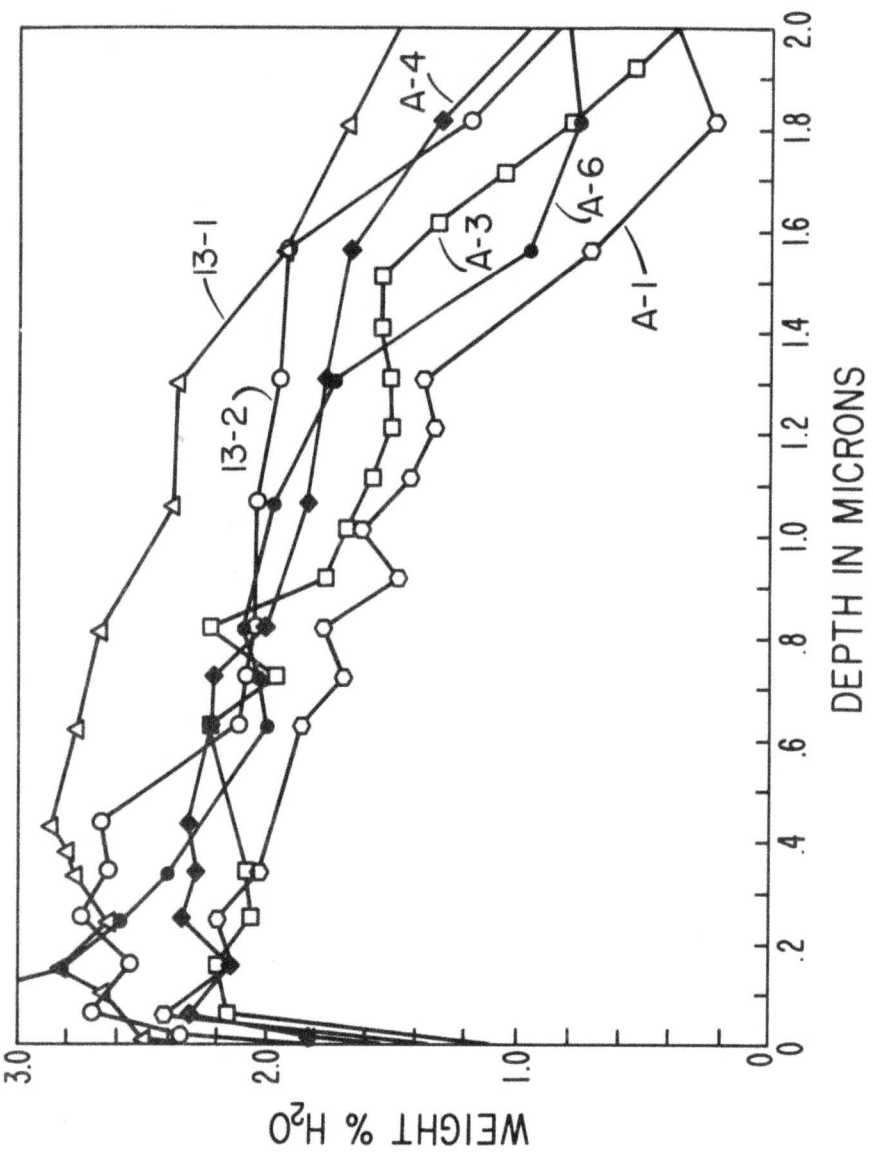

Plate 3-1. Water Profiles of Several Hydration Bands. The concentration at depth was measured at Cal Tech by nuclear reaction of fluorine with hydrogen. This data was provided by Professor T. A. Tombrello, Cal Tech. These curves indicate the form of the water concentration gradient in the hydrated layer. Several of these samples (A-3, A-4, A-6, and 13-2) indicate a two-step concentration gradient discussed by Ericson (1975). The A-series samples are obsidian artifacts from Amapa, Nayarit, Mexico (Meighan 1974), whose accession numbers and hydration values in microns are as follows: A-1, 246-1125B, 1.4; A-2, 246-3883H, 1.6; A-4, 246-3883I, 1.8; A-6, 246-3930, 2.1. Chemical analysis and other data on these samples is presented by Ericson and Berger (1976). The 13-series samples were measured on obsidian artifacts derived from the Borax Lake obsidian source; the provenience is unknown.

is not well understood at present, although on logical grounds it is expected to have an effect. It is suggested here that the second plateau is a slower and secondary reaction of the water and glass, possibly due to the out-diffusion of the alkalies. It is quite interesting to note that the activation energy of alkali ion migration in obsidian has been shown to be 18.7-19.6 Kcal/mole, using electrical conductivity measurements (Carron 1966:1666) which is a value equal to that shown for obsidian hydration (Friedman, Smith, and Long 1966). These observations suggest that the hydration process might best be described as a stepwise diffusion process (Ericson 1975). For the above reasons, the maximum water concentration in the hydration layer has been measured as an important intrinsic variable of the hydration process.

The measurement of the concentration of water in solid materials, particularly in ultrathin layers, has presented technical problems for physical science. A number of techniques, including outgassing, isotopic tracing, autoradiography, and infrared absorption spectrometry, have been employed with varying degrees of accuracy. Recent advances in nuclear technology have provided an accurate and efficient technique for water concentration measurement in ultrathin layers (Leich and Tombrello 1973; Leich, Tombrello, and Burnett, 1973).

The maximum water concentration was measured in the obsidian source samples using the following procedures. Samples were selected from among the obsidian collections at UCLA which had flat hydrated surfaces. A cylindrical core sample of $\frac{1}{2}$-inch diameter and 2 to 4 inches in length was removed from each sample using a $\frac{1}{2}$-inch diamond core drill. Two disk-shaped samples were cut from the core: (1) a sample with the original hydrated surface and (2) a sample from the interior, at least 1 inch from original hydrated surface. The samples were ground to obtain parallel surfaces with a thickness of less than than 1/8 inch. These hydrated and interior sample pairs were submitted to the Department of Physics, California Institute of Technology (Cal Tech).

At the nuclear facility at Cal Tech, groups of 11 samples and one standard were placed in the sample chamber of the linear accelerator, evacuated to a pressure of 10^{-6} Torr. The nuclear reaction $^1H\,(^{19}F,\,\alpha\gamma)\,^{16}O$ was used, where fluorine (^{19}F) nuclei at energies from 16-20 Mev obtained by linear acceleration, reacts with hydrogen nuclei (^1H) which produces oxygen (^{16}O), α-particles, and γ-radiation. The counts due to this reaction were measured at four different energies: 15.924 Mec, below resonance; 16.45 Mev, at resonance; 17.083 and 17.198 Mev, above resonance. The energies correspond to depths within the sample of 0.24, undet., 0.28, 0.33 microns. The average net number of counts from the sample due to the reaction was calculated by the formula presented in equation 3-5:

$$C_0 = \frac{C_4 + C_3}{2} - C_1 \qquad [3\text{-}5]$$

where:

C_0 = average net counts of ^1H in sample
C_1 = counts recorded at 15.924 Mev

C_3 = count recorded at 17.083 Mev
C_4 = counts recorded at 17.198 Mev

For each sample load a chlorite standard, containing 1.64 wt % hydrogen, was counted using a similar procedure. The weight percent of hydrogen for a particular sample was calculated using the formula presented in equation 3-6:

$$W_0 = W_c \frac{[dE/d(px)]_0 \, C_0}{[dE/d(px)]_c \, C_c} \qquad [3-6]$$

where:

W_0 = weight percent hydrogen in obsidian
W_c = weight percent hydrogen in chlorite standard = 1.64
$\frac{dE}{d(px)_0}$ = stopping power factor for obsidian = -9.6 Kev/cm^2-μg for hydrated obsidian; -9.4 Kev/cm^2-μg for unhydrated obsidian
$\frac{dE}{d(px)_c}$ = nuclear stopping power factor for chlorite = -9.2 Kev/cm^2-μg
C_0 = average net counts due to hydrogen in obsidian
C_c = average net counts due to hydrogen in chlorite

The conversion factor of 8.9364 was used to convert from weight percent hydrogen to weight percent water. The results of these measurements are reported in table 3-3.

Variable 12: Internal Water Concentration

The internal water concentration in obsidian is the residual water in hydrostatic equilibrium with the obsidian upon its extrusion. The role of water in the extrusion process is discussed by Ericson, MacKenzie, and Berger (1976). The actual concentration of water is a function of the water vapor pressure and temperature (Friedman, Long, and Smith 1963). In turn, the water vaporpressure is a function of the local hydrostatic pressure or depth within the extrusion, and the temperature a function of the extrusion temperature and thermodynamic cooling processes, discussed by Jaeger (1961). Considering the variability of the actual thermodynamic conditions, it is expected that the internal water concentration will be variable both between and within obsidian extrusions.

Recent papers have suggested that the internal water concentration within obsidian may constitute an important intrinsic variable of the hydration process (Kimberlin 1971, 1976; Lee et al. 1974; Berger and Ericson 1974). Kimberlin (1976) has suggested that there may be a lower limit of the water concentration (0.1 wt %), below which the hydration process is greatly retarded. This observation is supported by the unusual stability of tektite glasses relative to obsidians in the terrestrial environment. Although both glasses may have nearly identical chemical composition, their water concentrations differ greatly: 20 ppm and 10,000 ppm, respectively (O'Keefe 1964). Based on these comparisons, Berger and Ericson (1974) have suggested that the low concentration of internal water in tektites may account for their terrestrial stability. It appears that they entered the earth's atmosphere some

700,000 years ago, determined by potassium argon dating (Geiss and Hess 1958). Notably, Marshall (1961) suggests that the rarity of pre-Miocene obsidian ($>4.10^7$ years) may be attributed to the rapid process of hydration and subsequent devitrification.

Although the above studies are suggestive of a relationship between the variable and the hydration process, the degree and nature of the correlation remain to be determined. For these purposes, the internal water concentrations of obsidian samples were measured, using the procedures specified in the preceding section. The results of these measurements are reported in table 3-3.

Variables 13-14: The Structural Bonding of Water

Both the initial and final states of water bonding within the obsidian structure, variables 13 and 14, are expected to be important intrinsic variables of the hydration process. Firstly, it has been suggested that the hydration process is an autocatalytic process, having both diffusion and reaction components (Ericson, MacKenzie, and Berger 1976; Ericson 1975). It is expected that the reaction component would be a function of the nature of the structural sites available for water bonding in the hydration process. Secondly, the volumetric expansion, which produces the optical stress birefringence (Friedman and Smith 1960), should be directly related to the "relaxation" of the structure caused by the bonding of the water.

In obsidian, water is initially bound to the structure as bonded and free hydroxyl groups. A bonded hydroxyl group is bound to a bridging oxygen between two silicon ions. The linkage of the silicon-oxygen-silica groups forms the structural network of obsidian. A free hydroxyl group is bound to a nonbridging (or singly bound) oxygen which is itself bound to the structure by only one silicon ion.

The initial relationship between free and bonded hydroxyl groups in a glass is dependent upon the bonding condition of the singly bound oxygens (Scholze 1966:64). An extensive discussion on the expected relationships between the structure formed by a particular chemical composition and the water bonding in obsidian is presented by Ericson, MacKenzie, and Berger (1976). From this it is expected that excess alumina obsidians should show a predominance of the free hydroxyl groups relative to bonded hydroxyl groups, whereas excess alkali obsidians should show the opposite bonding characteristics. As an intrinsic variable of the hydration process, the initial state of water bonding within the structure would describe the ground state, i.e., how the internal water concentration is bound to the structure prior to the addition of water in the hydration process.

The final state of water bonding in the hydration layer (variable 14) would describe the result of the dynamic process of the reaction of water with the structure. Although both types of hydroxyl bonding produce increased density of the hydration layer, the formation of the bonded hydroxyl group should produce the greatest modification (relaxation) of the structural network producing the greatest volumetric expansion. It is the volumetric expansion of the hydration layer which produces the internal strain, leading to the appearance of the optical stress birefringence characterizaing the hydration layer

Table 3-4

PETROGRAPHIC STUDY OF CRYSTALLITES IN OBSIDIAN SOURCE SAMPLES

Source/Sample #	Percent Crystal- lization	Flow Banding	Feldspar Size	Feldspar Shape	Magnetite	Matrix	Misc.	Notes
Obsidian Butte								
1-1-1-1	4	X						holes in slide
1-1-2-5	2-3	X	M				B	brown streaks
1-1-3-1	25	X	M				B	
1-1-4-3	7	X	M				B	elongate structure
Coso								
2-1-1-17	10	X	MS		X	C	B	
Fish Springs								
3-1-1-11	1-2		S					clavalites
Inyo Craters								
4-1-0-6	90						Bblo	
4-2-0-3	100							andesite
4-3-0-								
4-5-1-								
4-5-2-1	80							
Mono Craters								
5-1-0-7	30		S	b	X		t	
5-2-0-1	5	X			X		mo	stringer
Mono Glass Mt.								
6-1-0-3	10	X	S		S	C	H	parallel bands
6-1-1-4	5	X	?	b			t	

(contd.)

Table 3-4 (contd.)

Source/Sample #	Percent Crystal-lization	Flow Banding	Feldspar Size	Feldspar Shape	Magnetite	Matrix	Misc.	Notes
West Queen								
7-1-0-10	4				X	VC		"cross" crystals
7-2-0-6	2		S					stippled texture
Duck Flat, Nev.								
8-1-0-								
Buck Mt.								
9-1-1-1	50–60		VS				B	tree-ring structure
9-1-1-5	3		S		X	C		
9-1-1-								
9-1-1-29	15	VX	S		X		B	hematite?
9-1-3-								
Cowhead Lake								
10-1-1-1	10		VS		X		T	
10-1-2-3	3		VS		X		B	
10-1-3-4								
10-2-1-1	17		S				BoTt	
10-2-2-1	20		?				B	fibrous
Sugarhill								
11-1-1-1	90		VS			H		
11-1-2-5	70	X	VS					
11-1-2-11	35	X	?	b			Tto	
11-1-3-7	98/S							
11-1-3-3	2	X				VC	b	

(contd.)

Table 3-4 (contd.)

Source/Sample #	Percent Crystal-lization	Flow Banding	Feldspar Size	Feldspar Shape	Magnetite	Matrix	Misc.	Notes
Modoc Glass Mt.								
12-1-0-1	90		LS		X		Tt	
12-2-0-9	1				X			
Medicine Lake								
12-3-0-1	70-80		LMS		X		Bbl	
Little Glass Mt.								
12-4-0-7	12		LS	Z		H		
Mt. Konocti								
13-1-0-5	1		LM			C		
13-2-0-1	20	X	LS				B	opaque stringer
13-3-0-3	10	X	LS				mB	
Borax Lake								
13-4-1-3	5	X	S		X			
13-4-2-1	3							lack of feldspars
Napa Glass Mt.								
14-1-1-1	30		S				mT	even zoning
14-1-2-1	2		S				B	numerous cracks
14-1-3-2	3-4	X	VS		X	C	B	stippled texture
14-2-0-4	5		X	?			0	stippled texture
14-3-0-1	10		S	E			0	many feldspars
14-4-0-3	4		S	E			0	crystals in zones
Annadel								
15-1-0-6	80	X	S	E	X			

(contd.)

Table 3-4 (cont.)

Source/Sample #	Percent Crystal- lization	Flow Banding	Feldspar Size	Feldspar Shape	Magnetite	Matrix	Misc.	Notes
Annadel continued 15-1-0-1	95	X	LS	Eb		sb	p	swirling
Monache Md. 16-1-0-1								
Beatty's Butte, Ore. 17-1-0-								
Long Valley, Nev. 18-1-0-								
Glass Mt., Ore. 19-1-0- 19-2-0-								
Glass Butte, Ore. 20-1-0-								
Rustler Canyon 21-0-0-0	25-30	X			X		T	"cross" crystals
Shoshone 22-1-0-12 22-3-0 22-4-0-11 22-4-0-	90-95 25	X	L L	Zb			0	tree bark structure
Jawbone Canyon 23-1-0-2	25	X	?	?	X			"scapulites" (contd.)

Table 3-4 (contd.)

Source/Sample #	Percent Crystal-lization	Flow Banding	Feldspar Size	Feldspar Shape	Magnetite	Matrix	Misc.	Notes
Casa Diablo								
24-1-1-1	17		S		X		mT	swirls
24-1-2-1	20	X	L		X			parallel magnetite
24-1-3-11	5	X	LS					polkitectic structure
Mt. Hicks, Nev.								
25-1-1-1/6	2-3		VS	b			0	opaques in "clouds"
Fletcher, Nev.								
26-1-1-1/8	0							almost nothing
Bodie Hills								
27-1-1-5	<1		?	s	X	C		
Pine Grove Hills, Nev.								
28-1-1-3	50		VS		X	VC	mo	no large crystals

Key to Table 3-4

X = present
L = large (1000μ-2mm)
M = medium (100μ-1000μ)
S = small (100μ-0μ)
l = lathe
E = enhedral
a = anhedral
s = subhedral
C = clear
V = very

B = belonites*
b = bacillite*
T = acicular trichites*
t = asteroidal trichites*
o = opaques
p = perlitic cracks
z = zoned feldspar
m = many
H = homogeneous
* (Terminology after Clark 1961a, fig. 14).

(Friedman and Smith 1960). Therefore, among all the intrinsic variables under consideration, it is expected that the variables which promote the formation of <u>bonded</u> hydroxyl groups will correlate with rapid rates of hydration in obsidian.

Variable 15: Degree of Crystallinity

The degree of crystallinity within the obsidian samples, chosen as an intrinsic variable, gives a relative measure of the effects of enrichment or depletion of certain elements in the glass phase of obsidian. Since it is the chemical composition of the glass phase that is thought to influence the hydration process, the chemical analyses of major elements, variables 1-8, do not take into consideration the effects of elementary changes of the bulk glass due to crystallization. The degree of crystallinity in the obsidian samples, tabulated in tables 3-3 and 3-4, is an attempt to control this secondary variation. In addition, there are two other side benefits in knowing the degree of crystallinity. One, the degree of crystallinity should affect its lithic qualities for stone tool manufacture. Second, the crystallinity may be shown to be a means to characterize the obsidian to specific sources. Table 3-4 suggests that many obsidian sources have the potential for this form of characterization.

The degree of crystallinity, expressed as a percent, was estimated by comparing the amount of crystallization in individual samples. Samples were prepared courtesy of UCLA Museum of Geology as petrographic thin sections, which were standardized at 30 microns thickness. The samples were examined with a standard binocular petrographic microscope at magnifications of 40x, 100x, and 400x. This examination was conducted by Victoria Bennett, UCLA Obsidian Hydration Dating Laboratory. The estimation of the degree of crystallinity was made with standard percent determination charts.

A petrographic analysis of the obsidian source samples was conducted at the same time and is reported in table 3-4. Minerals and textures were determined using standard optical microscopy and are described in petrographic terminology (Kerr 1959). The crystallites or microlites were described using the terminology defined by Clark (1961a:Fig. 14). Early geologists suggested that crystallites may represent a stage between the amorphous state and the crystalline (Johannsen 1939). Recent technological developments of electron microscopes, electron probes, and ionic microscopes may provide the means to resolve this interesting hypothesis. If true, the states of matter, rather than being portioned into gas, liquid, and solid, would become a continuum of matter. This redefinition of a fundamental concept of the physical sciences would be quite significant. An examination of the crystallites in obsidian holds that exciting potential.

CALCULATED INTRINSIC VARIABLES

The calculated intrinsic variables in table 3-5 were derived directly from the empirical variables. They are the second set of variables considered to influence the rates of hydration among the obsidian sources. These variables have been derived as a result of several research concerns, based on their expected theoretical relationships to the hydration phenomena:

1. If an empirical variable is significantly influenced by a second empirical variable, then the first variable is corrected by the second variable.

2. If an empirical variable is a function of two independent variables, one of which is deemed significant, then the empirical variable is decoupled from the insignificant variable.

3. If an empirical variable is expected to be one of a set of interrelated empirical variables, then the nature of this interrelationship is described.

In each case, the specific calculations and details of variable selection are specified in the sections which follow. They are, in their order of presentation, the following: (variable 16) silicon-oxygen ratio, S; (variable 17) oxygen activity, O; (variable 18) structural factor, R; (variable 19) alumina factor, A; (variable 20) specific volume, Ω.

Variables 16-18: Silicon-Oxygen Ratio, Oxygen Activity, and Structural Factor

The silicon-oxygen ratio, variable 16, is a measure of the amount of modification of the silica network, which is modified by the addition of other cations other than silicon. In silica glass, an unmodified glass, the silicon-oxygen ratio is equal to 0.5. The progressive modification of the silica network by other cations reduces the silicon-oxygen ratio to less than 0.5. The significance of the relationship of this variable to the hydration process is discussed by Ericson, MacKenzie, and Berger (1976). Ericson and Berger (1976) suggest that there is a high correlation between the silicon-oxygen ratio and the hydration process in two case studies. The silicon-oxygen ratio is defined in equation 3-7 and a sample calculation is derived in table 3-6 (Huggins 1944).

$$S = \frac{\text{silicon (moles)}}{\text{oxygen (moles)}} \qquad [3\text{-}7]$$

Carron (1969) suggests the importance of the oxygen activity, variable 17, and the number of broken structural bonds as structural factor R, variable 18, in determining the viscosity of glasses, which he related to the number of nonbridging oxygen atoms. As previously stated, the nonbridging oxygen with the structure provide potential sites for <u>free</u> hydroxyl groups. The number of active oxygens, O_A, and the structural factor, R, are defined in equations 3-8 and 3-9 (Carron 1969:table 2):

$$O_A = 2\,(O - 2T) \qquad [3\text{-}8]$$

where:

O_A = number of active oxygens
O = number of oxygen atoms corresponding to tetrahedrally coordinated and interstitial cations

Table 3-5

CALCULATED INTRINSIC VARIABLES

Source/Sample#	Silicon Oxygen Factor, S	Oxygen Activity Factor, O	Structural Factor, R	Aluminum Factor, A	Calculated Density (g/cc)	Specific Volume Factor	$\dfrac{Na}{Na+K}$
Obsidian Butte							
1-1-1-1	0.4203	10.0	6.6	0.357	2.4011	x	0.600
1-1-2-5	0.4198	9.8	6.4	0.702	2.4030	x	0.590
1-1-3-1	0.4182	8.5	5.6	1.111	2.4053	x	0.586
1-1-4-3	0.4233	7.6	4.9	0.407	2.3916	x	0.565
Coso							
2-1-1-17	0.4254	2.5	1.6	1.083	2.3765	100.79	0.549
Fish Springs							
3-1-1-11	0.4245	1.4	0.9	0.976	2.3750	100.45	0.553
Inyo Craters							
4-1-0-6	0.4109	6.9	4.6	2.024	2.4164	104.35	0.538
4-2-0-3	0.4124	7.7	5.1	1.441	2.4140	99.04	0.514
4-3-0-2	0.4028	14.2	9.6	1.861	2.4450	99.31	0.556
4-5-1-2	0.4185	6.4	4.2	0.754	2.3978	103.05	0.514
4-5-2-1	0.4177	6.5	4.3	0.788	2.4000	102.39	0.518
Mono Craters							
5-1-0-7	0.4262	3.0	1.9	1.009	2.3755	101.36	0.515
5-2-0-1	0.4246	3.0	1.9	0.898	2.3791	100.69	0.544
Mono Glass Mt.							
6-1-0-3	0.4262	1.4	0.9	1.114	2.3712	97.67	0.520
6-1-1-4	0.4269	1.8	1.2	0.751	2.3706	x	0.507

(contd.)

Table 3-5 (contd.)

Source/Sample #	Silicon Oxygen Factor, S	Oxygen Activity Factor, O	Structural Factor, R	Aluminum Factor, A	Calculated Density (g/cc)	Specific Volume Factor	$\dfrac{Na}{Na + K}$
West Queen							
7-1-0-10	0.4240	2.7	1.7	0.994	2.3772	100.85	0.542
7-2-0-6	0.4216	1.5	1.0	1.224	2.3821	101.72	0.537
Duck Flat, Nev.							
8-1-0-5	0.4274	7.0	4.6	0.148	2.3805	100.94	0.554
Buck Mt.							
9-1-1-1	0.4238	2.5	1.6	1.251	2.3775	99.76	0.523
9-1-1-5	0.4227	1.9	1.3	1.611	2.3794	97.89	0.516
9-1-1-13	0.4231	1.8	1.1	1.558	2.3785	100.64	0.516
9-1-1-29	0.4234	1.0	0.6	1.744	2.3773	98.99	0.511
9-1-3-3	0.4231	2.8	1.8	1.374	2.3799	100.50	0.528
Cowhead Lake							
10-1-1-1	0.4261	0.7	0.4	1.104	2.3710	101.21	0.582
10-1-2-3	0.4261	0.8	0.5	1.005	2.3710	99.38	0.614
10-1-3-4	0.4259	1.9	1.3	0.701	2.3724	x	0.606
10-2-1-1	0.4261	1.6	1.0	0.779	2.3717	100.16	0.595
10-2-2-1	0.4264	0.7	0.5	1.049	2.3705	101.08	0.580
Sugarhill							
11-1-1-1	0.4232	3.6	2.3	1.301	2.3820	98.98	0.503
11-1-2-5	0.4234	3.3	2.2	1.339	2.3813	100.20	0.488
11-1-2-11	0.4242	5.8	3.9	1.524	2.3768	98.66	0.586
11-1-3-3	0.4255	4.4	2.9	0.540	2.3763	100.82	0.518
11-1-3-7	0.4226	3.4	2.2	1.527	2.3826	96.60	0.500

(contd.)

Table 3-5 (contd.)

Source/Sample #	Silicon Oxygen Factor, S	Oxygen Activity Factor, O	Structural Factor, R	Aluminum Factor, A	Calculated Density (g/cc)	Specific Volume Factor	$\frac{Na}{Na + K}$
Modoc Glass Mt.							
12-1-0-1	0.4139	9.5	6.3	1.874	2.4125	99.75	0.569
12-2-0-9	0.4158	6.9	4.5	2.127	2.4056	100.10	0.546
Medicine Lake							
12-3-0-1	0.3958	20.4	13.9	3.829	2.4371	102.01	0.598
Little Glass Mt.							
12-4-0-7	0.4121	5.8	3.9	2.202	2.4132	102.86	0.586
Mt. Konocti							
13-1-0-5	0.4190	6.1	4.0	1.804	2.3953	101.64	0.518
13-2-0-1	0.4189	5.3	3.5	1.970	2.3948	99.87	0.496
13-3-0-3	0.4189	5.6	3.7	1.797	2.3952	100.84	0.503
Borax Lake							
13-4-1-3	0.4268	4.1	2.7	0.929	2.3751	106.16	0.498
13-4-2-1	0.4240	3.3	2.1	1.451	2.3797	100.97	0.514
Napa Glass Mt.							
14-1-1-1	0.4237	4.0	2.6	0.824	2.3819	100.16	0.543
14-1-2-1	0.4252	3.8	2.5	0.730	2.3791	100.58	0.579
14-1-3-2	0.4136	-2.7	-1.7	2.705	2.3970	99.57	0.616
14-2-0-4	0.4239	3.2	2.1	0.783	2.3813	100.46	0.585
14-3-0-1	0.4251	4.4	2.8	0.468	2.3797	100.11	0.586
14-4-0-3	0.4251	4.2	2.8	0.521	2.3800	100.61	0.573
Annadel							
15-1-0-1	0.4173	3.3	2.1	2.656	2.4001	x	0.497
15-1-0-6	0.4173	7.7	5.1	1.374	2.4053	98.49	0.654

(contd.)

Table 3-5 (contd.)

Source/Sample #	Silicon Oxygen Factor, S	Oxygen Activity Factor, O	Structural Factor, R	Aluminum Factor, A	Calculated Density (g/cc)	Specific Volume Factor	$\dfrac{Na}{Na + K}$
Monache Md.							
16-1-0-1	0.4211	-0.9	-0.6	1.909	2.3795	x	0.569
Beatty's Butte, Ore.							
17-1-0-4	0.4094	0.2	0.1	3.084	2.4093	x	0.552
Long Valley, Nev.							
18-1-0-1	0.4265	9.2	6.1	-0.396	2.3869	x	0.546
Glass Mt., Ore.							
19-1-0-6	0.4170	8.3	5.5	0.825	2.4080	x	0.572
19-2-0-2	0.4203	10.4	6.9	0.040	2.4027	x	0.570
Glass Butte, Ore.							
20-1-0-1	0.4263	0.74	0.5	1.352	2.3706	x	0.555
Rustler Canyon							
21-1-0-0	0.4253	5.8	3.8	-0.021	2.3802	x	0.544
Shoshone							
22-1-0-12	0.4207	4.7	3.1	1.270	2.3885	x	0.477
22-3-0-5	0.4207	4.8	3.1	2.153	2.3894	x	0.527
22-4-0-11	0.4229	3.0	2.0	1.576	2.3799	x	0.486
22-4-0-16	0.4177	-1.1	-0.7	2.510	2.3878	x	0.551
Jawbone Canyon							
23-1-0-2	0.4174	-3.5	-2.6	2.989	2.3862	x	0.485
Casa Diablo							
24-1-1-1	0.4198	4.0	2.6	1.539	2.3913	x	0.468

(contd.)

Table 3-5 (contd.)

Source/Sample #	Silicon Oxygen Factor, S	Oxygen Activity Factor, O	Structural Factor, R	Aluminum Factor, A	Calculated Density (g/cc)	Specific Volume Factor	$\frac{Na}{Na+K}$
Casa Diablo continued							
24-1-2-1	0.4218	4.0	2.6	1.094	2.3870	x	0.494
24-1-3-11	0.4213	4.6	3.0	1.222	2.3887	x	0.472
Mt. Hicks, Nev.							
25-1-1-1/6	0.4277	0.2	0.1	1.260	2.3661	x	0.502
Fletcher, Nev.							
26-1-1-1/8	0.4245	2.1	1.3	1.043	2.3758	x	0.522
Bodie Hills							
27-1-1-5	0.4241	1.7	1.1	1.212	2.3762	x	0.508
Pine Grove Hills, Nev.							
28-1-1-3	0.4264	2.0	1.3	1.076	2.3716	x	0.489

T = member of tetrahedrally coordinated cations = $Si^{+4} + Al^{+3} + Fe^{+2} + P^{+5}$

$$R = 100 \left(\frac{O_A}{T}\right) \quad\quad [3-9]$$

where:

R = structural factor of the percentage of broken oxygen bonds
O_A = number of active oxygens
$T = Si^{+4} + Al^{+3} + Fe^{+3} + P^{+5}$

A sample calculation of both variables 17-18 are derived in table 3-7. The results of the calculations of variables 16-18 are presented in table 3-5.

Variable 19: Alumina Factor

The alumina factor, A, is a measure of the amount of alumina relative to alkalies that are present within the glass. This quantity may be related to the characteristic type of water bonding with the obsidian.

Scholze (1959), using infrared absorption analysis, has shown that excess alumina increases the number of free hydroxyl groups relative to bonded hydroxyl groups. In glasses, where the ratio of alumina and alkalies was 1:1, only bonded hydroxyl groups were present, the characteristic of silica glass. In glasses where the alkalies exceeded alumina, bonded hydroxyl groups were more prevalent. The nature of these relationships is presented as follows:

1. If A > 1, free > bonded hydroxyl groups.
2. If A = 1, bonded hydroxyl groups.
3. If A < 1, bonded < free hydroxyl groups.

If the characteristic bonding of water plays a fundamental role in influencing the hydration process, then the factor A might be an important variable.

The factor A was calculated for individual obsidian samples, based upon the chemical analysis. A sample calculation of this factor is presented in table 3-8. The results of the calculations are presented in table 3-5.

Variable 20: Specific Volume

The specific volume, Ω (Ericson, MacKenzie, and Berger 1976), is a measure of volume of the measured density and calculated density based on chemical composition (Huggins 1944). As discussed in the section entitled "Variable 9: Density," density is a composite physical property of two independent variables, namely, volume and mass. The specific volume is a factor normalized and independent of the effects of mass among samples. Therefore, this factor describes the state of the volume of the obsidian, i.e., whether or not the annealed (or minimum) volume has been achieved. In this regard, Bruchner (1965) has shown that the thermal history affects the diffusion of water in commercial glasses. This finding suggests that the specific volume as a measure of obsidian might be an important variable of the hydration process (Ericson and Berger 1976).

Table 3-6

SAMPLE CALCULATION OF SILICON-TO-OXYGEN RATIO,
FACTOR S (after Huggins 1944)

Column	(1)	(2)	(3)	(4)
Oxide	Sample (1-1-1-1) Normalized Wt. %	Atomic Weight Oxide	Number Oxygen	Oxygen Factor
SiO_2	75.594	60.0848	2.0	2.51622
Al_2O_3	12.258	101.9612	1.5	0.18032
CaO	0.751	56.0794	1.0	0.01339
Na_2O	4.270	61.9790	0.5	0.03445
K_2O	4.324	94.2034	0.5	0.02947
Fe_2O_3	2.516	159.6922	1.5	0.02363
MgO	0.132	40.3114	1.0	0.00327
TiO_2	0.155	79.8988	2.0	0.00388
Total	100.0			2.79674

(Col. 4) = (Col. 1) (Col. 3) ÷ (Col. 2)

$$\text{Fractional of silicon} = \frac{\text{Atomic Weight Si}}{\text{Atomic Weight SiO}_2} = \frac{28.085}{60.0848} = 0.4674$$

Silicon fraction of oxygen factor = 0.4674 (2.51622) = Factor S =

$$\frac{\text{silicon fraction of oxygen factor}}{\text{total oxygen factor}} = \frac{(0.4674)\,(2.51622)}{2.79674}$$

$\underline{S = 0.4203}$

The specific volume is calculated by the formula described in equation 3-10 (Ericson, MacKenzie, and Berger 1976):

$$\Omega = 100\left[1 + \frac{p_c - p_m}{p_c}\right] \qquad [3\text{-}10]$$

where:

Ω = specific volume
p_c = calculated density (after Huggins 1944)
p_m = measured density

The relationship between the annealed volume and specific volume of a particular sample is evaluated by the following criteria:

Table 3-7

SAMPLE CALCULATION OF NORMALIZED ANALYSIS (1-1-1-1)
ACTIVITY FACTOR, O_A, AND THE NUMBER OF BROKEN
STRUCTURAL BONDS, R
(after Carron 1969:table 2)

Weight Percent Oxides	Number of gram-atoms in 10,000 grams		
	Tetrahedral Cations ("glass formers")	Interstitial Cations ("glass modifiers")	Corresponding Oxygen Atoms
SiO_2: 75.594	Si^{+4}: 125.81		251.62
Al_2O_3: 12.258	Al^{+3}: 24.04		36.06
Fe_2O_3*: 1.258	Fe^{+3}: 1.575		2.36
FeO*: 1.258		Fe^{+2}: 1.75	1.75
MnO: 0.04		Mn^{+2}: 0.06	0.06
MgO: 0.132		Mg^{+2}: .33	0.33
CaO: 0.751		Ca^{+2}: 1.34	1.34
Na_2O: 4.270		Na^+: 13.78	6.89
K_2O: 4.324		K^+: 9.18	4.59
TiO_2: 0.155		Ti^{+4}: .19	0.39
Total: 100.04	T = 151.425		O = 305.39

$O_A = 2 \times (O - 2T) = 2 \times (305.39 - 302.85) = 3.08$

$R = 100 \frac{OA}{T} = \frac{100 \ (5.08)}{151.425} = 3.35$

*Fe_2O_3/FeO treated as 1.

---

1. $\Omega > 100$ (specific volume = unannealed volume)
2. $\Omega = 100$ (specific volume = annealed volume)
3. $\Omega < 100$ (specific volume = annealed volume + additional densification due to crystallization)

The specific volume, factor Ω, was calculated for individual samples, using the format of the sample calculation which is presented in table 3-9. The results of the sodium-potassium calculations are presented in table 3-10.

DETERMINATION OF THE INTRINSIC VARIABLES OF THE HYDRATION PROCESS

The preceding sections have demonstrated the variation of the hydration rates among obsidian sources. The hypothesis presented by Ericson and

Table 3-8

SAMPLE CALCULATION OF FACTOR A, ALUMINA FACTOR

Column:	(1) Sample (1-1-1-1) Normalized Wt. %	(2) Atomic Weight Oxide	(3) Mole Factor	(4) Mole %	(5) Calculations
SiO_2	75.594	60.0848	1.2579	82.361	(Col. 3) = (Col. 1) ÷ (Col. 2)
Al_2O_3	12.258	101.9612	0.1203	7.874	(Col. 4) = 100 (Col. 3) ÷ Total (Col. 3)
CaO	0.751	56.0794	0.0134	0.877	
Na_2O	4.270	61.9790	0.0689	4.510	Factor A = mole % Al_2O_3 - (mole % Na_2O + mole % K_2O)
K_2O	4.324	94.2032	0.0459	3.007	
Fe_2O_3	2.516	159.6922	0.0158	1.031	= 0.357 mole % Al_2O_3
MgO	0.132	40.3114	0.0033	0.214	
TiO_2	0.155	79.8988	0.0019	0.127	
Total	100	---	1.5273	100	

Table 3-9

SAMPLE CALCULATION OF FACTOR Ω, SPECIFIC VOLUME FACTOR

Column:	(1)	(2)	(3)	(4)
Oxide	Sample (2-1-1-17) Normalized Wt. %	Density Factor (Huggins 1944)	Density Components	Calculations
SiO_2	77.849	0.4409	.3432	Col. 3 = (Col. 1) x (Col. 2) ÷ 100
Al_2O_3	12.322	0.373	.0460	Calculated density - ρ_c = 1 ÷ Total (Col. 4) = 2.3765
CaO	0.488	0.231	.0011	
Na_2O	3.547	0.324	.0115	Empirical density = ρ_m = 2.3576
K_2O	4.425	0.357	.0158	$\Omega = 100 \left[1 + \dfrac{\rho_c - \rho_m}{\rho_c} \right] = 100 \left[1 + \dfrac{2.3765 - 2.3576}{3.765} \right]$
Fe_2O_3	1.228	0.225	0.0028	
MgO	0.088	0.322	0.0003	$\Omega = 100.79$
TiO_2	0.053	0.243	0.0001	
Total L	100		0.42079	

Table 3-10

SAMPLE CALCULATION OF Na/Na+K

Column	(1)	(2)	(3)	(4)
Oxide	Sample (1-1-1-1) Normalized Wt. %	Atomic Weight Oxide	Mole Factor	Calculations
Na_2O	4.270	61.9790	Na: 0.0689	$\frac{Na}{Na+K} = \frac{0.0689}{0.1147949} = 0.60$
K_2O	4.324	94.2034	K: 0.0459	

Berger (1976) expects that the intrinsic properties of individual obsidian sources will account for this variation. The isolation of the major intrinsic variables of the hydration process will lead to a greater understanding of the hydration process, a very complex phenomenon.

Stepwise multiple regression analysis was selected as the multivariate statistical technique to examine this problem. This analytical technique allows one to isolate the determiners of hydration rate variation and to specify how effective these variables are as determiners. In this analysis, a stepwise process is used to enter each variable in sequence of its explained variability. As the number of variables increases, a complex descriptive ("explanatory") network of variation is developed.

Fortunately for this analysis, the previous discussion has placed each variable within a certain theoretical framework related to the interaction of the internal structure and the hydration process. A more detailed discussion is presented in Ericson, MacKenzie, and Berger (1976). Both the theoretical framework and the results of statistical analysis will allow for an evaluation of the results.

Both the empirical and calculated intrinsic variables for each obsidian source, presented in table 3-11, were considered as independent variables in regression analysis. The activation energies of the induced hydration experiments, presented in table 2-6, were used as the dependent variable in the regression analysis. These experimental values appear to be more internally consistent relative to the empirical or archaeologically derived values. The computer program, BMDP2R, was used to perform stepwise multiple regression analysis. In this final analysis, because all data were not present to fill all cells of the variable matrix, several variables were excluded. These variables are the density, hardness, percent crystallinity, and specific volume. The results of statistical analysis are presented in table 3-12.

DISCUSSION

The intrinsic variables of the hydration process are definitely linked to the internal structure of the glass. Since glasses are high entropy materials, a determination of their internal structure is quite difficult. Water can be used as a probe to determine certain aspects of the structure. The results of multivariate analysis, presented in table 3-12, have to be interpreted on a structural basis.

First, it is important to note that there are many variables which can alter the hydration rate in obsidian. Without a set of major known variables, it will be difficult to predict a given rate of physical and chemical data. In fact it is the state of physical and chemical properties of a particular source which governs the variability of the hydration rate for that source. Second, the analysis on table 3-12 is presented here as a heuristic device, not a valid statistical test, due to the inadequate sample size (N=10). Further analysis will be conducted to resolve the variables of the hydration process on an expanded data set.

Table 3-11

SUMMARY TABLE OF CHEMICAL AND PHYSICAL PROPERTIES AND
CALCULATED FACTORS FOR OBSIDIAN SOURCE SAMPLES

	N	SiO_2	Al_2O_3	CaO	Na_2O	K_2O	Fe_2O_3	MgO	TiO_2	Density	Hardness	Percent Crystal
Obsidian Butte	4	73.95	12.13	0.71	3.97	4.27	2.50	0.11	0.13	-	562	10
Coso	1	78.37	12.40	0.49	3.57	4.45	1.24	0.09	0.05	2.36	773	10
Fish Springs	1	78.40	12.88	0.51	3.81	4.69	0.81	0.04	0.04	2.36	714	2
Inyo Craters	5	69.83	13.70	1.04	3.75	5.11	2.22	0.19	0.20	2.38	-	80
Mono Craters	2	78.31	12.28	0.54	3.47	4.68	1.30	0.04	0.05	2.35	756	18
Mono Glass Mt.	2	77.92	12.09	0.44	3.32	4.78	0.84	0.06	0.05	2.43	770	8
W. Queen	2	75.03	12.69	0.55	3.61	4.68	0.91	0.91	0.08	2.35	746	2
Duck Flat, Nev.	1	82.26	11.63	0.31	3.83	4.69	2.14	0.05	0.11	2.37	832	-
Buck Mt.	5	79.14	13.10	0.90	3.38	4.73	0.88	0.12	0.11	2.39	705	24
Cowhead Lake	5	77.74	12.34	0.31	3.94	4.07	0.87	0.05	0.03	2.36	698	13
Sugarhill	5	79.38	12.81	0.75	3.29	4.99	1.19	0.12	0.11	2.40	601	59
Modoc Glass Mt.	2	73.25	13.42	1.58	3.51	4.23	2.10	0.25	0.27	2.41	742	45
Mt. Konocti	3	76.40	13.21	1.26	3.16	4.70	1.66	0.23	0.24	2.38	541	10
Borax Lake	2	81.23	12.45	0.81	3.23	4.79	1.29	0.16	0.08	2.30	734	4
Napa Glass Mt.	6	75.87	12.75	0.35	3.97	4.36	1.45	0.04	0.07	2.38	789	9
Annadel	2	73.38	13.33	1.07	3.66	3.96	2.36	0.10	0.15	2.44	590	88
Monache Mt., Ore.	1	77.42	14.08	0.72	3.82	4.40	0.70	0.07	0.03	-	-	-
Beatty's Butte	1	63.80	14.58	0.93	3.49	4.29	1.39	0.13	0.12	-	-	-
Long Valley, Nev.	1	75.62	10.70	0.24	3.75	4.75	2.35	0.00	0.09	-	-	-
Glass Mt., Ore.	2	72.20	12.42	0.46	4.09	4.67	2.69	0.05	0.18	-	-	-

(contd.)

Table 3-11 (contd.)

	N	SiO_2	Al_2O_3	CaO	Na_2O	K_2O	Fe_2O_3	MgO	TiO_2	Density	Hardness	Percent Crystal
Glass Butte, Ore.	1	79.69	12.51	0.49	3.50	4.62	0.91	0.08	0.06	–	–	–
Rustler Canyon	1	76.82	11.78	0.20	3.90	4.98	1.45	0.06	0.13	–	689	27
Shoshone	4	72.61	12.62	0.92	3.05	4.45	1.16	0.23	0.11	–	572	60
Jawbone Canyon	1	63.67	12.93	0.51	2.65	4.29	0.86	0.07	0.03	–	546	25
Casa Diablo	3	74.37	12.73	0.75	3.13	5.19	1.50	0.09	0.14	–	799	14
Mt. Hicks, Nev.	1	83.40	12.73	0.48	3.24	4.90	0.72	0.06	0.06	–	874	3
Fletcher, Nev.	1	75.90	12.29	0.66	3.38	4.73	0.79	0.05	0.07	–	793	0
Bodie Hills	1	73.16	12.01	0.63	3.15	4.64	0.78	0.09	0.07	–	907	0
Pine Grove Hills, Nev.	1	77.27	11.86	0.75	3.03	4.81	0.73	0.05	0.06	–	671	50

Table 3-11 (contd.)

	Max. Water	Internal Water	S	O	R	A	Cal. Density	Specific Value	$\frac{Na}{Na+K}$
Obsidian Butte	2.66	0.85	0.420	9.0	5.9	0.64	2.40	–	0.585
Coso	5.95	0.98	0.425	2.5	1.6	1.08	2.38	100.8	0.549
Fish Springs	4.09	0.57	0,425	1.4	0.9	0.98	2.38	100.5	0.553
Inyo Craters	3.33	0.98	0.412	8.3	5.6	1.37	2.42	101.6	0.528
Mono Craters	3.72	0.89	0.425	3.0	1.9	0.95	2.38	101.0	0.530
Mono Glass Mt.	3.63	0.61	0.426	1.6	1.1	0.93	2.37	97.7	0.514
W. Queen	3.60	0.70	0.423	2.1	1.4	1.11	2.38	101.3	0.540
Duck Flat, Nev.	3.25	0.78	0.427	7.0	4.6	0.15	2.38	100.9	0.554
Buck Mt.	3.47	0.53	0.423	2.0	1.3	1.51	2.38	99.6	0.519
Cowhead Lake	4.09	0.90	0.426	1.1	0.7	0.93	2.37	100.5	0.595
Sugarhill	3.90	0.60	0.424	4.1	2.7	1.25	2.38	99.1	0.519
Modoc Glass Mt.	4.19	0.81	0.415	8.2	5.4	2.00	2.41	99.9	0.558
Mt. Konocti	4.40	0.89	0.419	5.7	3.7	1.86	2.39	100.8	0.592
Borax Lake	3.86	0.74	0.425	3.7	2.4	1.19	2.38	103.6	0.506
Napa Glass Mt.	4.21	0.83	0.423	2.8	1.9	1.01	2.38	100.2	0.580
Annadel	6.68	0.33	0.417	5.5	3.6	2.02	2.40	98.5	0.576
Monache Mt., Ore.	6.87	3.90	0.421	-1	-0.6	1.91	2.38	--	0.569
Beatty's Butte	5.05	0.80	0.409	0.2	0.1	3.08	2.41	--	0.552
Long Valley, Nev.	6.47	0.65	0.427	9.2	6.1	-0.40	2.39	--	0.546
Glass Mt., Ore.	4.05	1.62	0.419	9.4	6.2	0.43	2.40	--	0.571

Table 3-11 (contd.)

	Max. Water	Internal Water	S	O	R	A	Cal. Density	Specific Value	$\dfrac{Na}{Na+K}$
Glass Butte, Ore.	3.34	0.52	0.426	0.7	0.5	1.35	2.37	–	0.555
Rustler Canyon	2.14	0.86	0.425	5.8	3.8	-0.02	2.38	–	0.544
Shoshone	3.98	5.37	0.421	2.9	1.9	1.88	2.39	–	0.510
Jawbone Canyon	5.90	5.59	0.417	-3.5	-2.6	2.99	2.39	–	0.485
Casa Diablo	2.83	0.89	0.421	4.2	1.3	1.29	2.39	–	0.478
Mt. Hicks, Nev.	4.04	0.52	0.428	0.2	0.1	1.26	2.37	–	0.502
Fletcher, Nev.	3.75	1.31	0.425	2.1	1.3	1.04	2.38	–	0.522
Bodie Hills	3.09	0.65	0.424	1.7	1.1	1.21	2.38	–	0.508
Pine Grove Hills, Nev.	4.55	0.68	0.426	2.0	1.3	1.08	2.37	–	0.489

Table 3-12

VARIABLES OF THE HYDRATION PROCESS

Variable	Mean	Standard Deviation	Partial Correlation	Step	Stepwise Multiple R
SiO_2	77.38	3.17	0.4205	5	0.944
Al_2O_3	12.77	0.51	-0.3387	-	-
CaO	0.80	0.35	-0.5824	1	0.582
Na_2O	3.51	0.45	0.0186	-	-
K_2O	4.50	0.34	0.2249	-	-
Fe_2O_3	1.40	0.61	-0.2959	-	-
MgO	0.12	0.07	-0.5515	-	-
TiO_2	0.11	0.08	-0.2852	2	0.808
Final H_2O	4.27	1.14	-.3790	-	-
Initial H_2O	0.67	0.27	0.3578	4	0.916
S Factor	0.423	0.003	0.4952	-	-
Density	2.38	0.03	-0.2417	3	0.891
Hardness			0.26657	6	0.984
A Factor	1.48	0.59	-0.5610	-	-
Cal. Density	2.39	0.59	-0.3274	-	-
Crystallization	11.80	24.18	-0.335	-	-
Activation Energy	26.80	7.49	-	-	-

PART 2

REGIONAL EXCHANGE SYSTEMS

Behind the mask is a man; behind ceremonies are resources.

Chapter 4

INTRODUCTION TO THE STUDY OF REGIONAL EXCHANGE SYSTEMS

INTRODUCTION

The overall objective of the second part of this monograph is to define and operationalize a methodology which will facilitate the study of regional exchange systems. The object is not to resolve specific models but to look at a set of data to gain a basic understanding of regional exchange. The utility of the methodology will be shown through the analysis of a selected data set. This will establish that several variables can be examined in a space utility framework, a possibility which until now has not been well examined. In fact, the study of regional exchange systems represents a definite challenge. The author recently contributed several papers to and coedited a published volume with Professor T. K. Earle entitled <u>Exchange Systems in Prehistory</u> (Earle and Ericson 1977). The aim of this volume was to revitalize interest in (see review by Webb 1975) and explore new methodological avenues for researching exchange systems. Many aspects of exchange not dealt with here are covered in the volume and the interested reader may refer to it directly.

The analysis presented here represents a serious examination of the prehistoric exchange systems in California. The analysis utilizes the results presented in part 1, where a number of highly technical processes and tests proper to the physical sciences, rather than anthropology, are discussed. Certain aspects of obsidian characterization and hydration dating are crucial to the research problem investigated in this part of the dissertation.

California has been selected as the region of study since it incorporates a high diversity of prehistoric cultures, environment, and natural biological resources. These will eventually permit the study of the effects of different variables influencing the exchange of items. Of particular interest to this study are the number of obsidian sources located in California used as sources for raw material for chipped stone tool manufacture. The obsidian was exchanged over fairly long distances. Obsidian can thus be used as a "tracer" for studying the internal structure of a prehistoric exchange system in a manner similar to the use of radioactive isotopes in nuclear medicine.

The geographical distribution of obsidian, originating from a specific source, which is the consequence of the processes of exchange and utilization, and its subsequent deposition in the archaeological record, is referred to in this study as a <u>regional exchange system</u>. It is important to note that the spatial distribution of the obsidian is not a system, but rather it is the consequence of a system. Nevertheless, for our purposes if we are to study regional exchange systems, we must treat the spatial distributions of exchange items as systems, which will allow us to study their structural characteristics, internal organization, diachronic stability, and cultural interrelationships.

The study of regional exchange systems in California is a challenge and requires the application of a set of interdisciplinary techniques. There are five methodological problems which have to be surmounted. First, it will be necessary to expand Binford's notion of space utility. Second, it will be necessary to describe the prehistoric distribution of obsidian in California. Third, since there are several obsidian sources, it will be necessary to distinguish distributions of specific sources of obsidian, and to describe the characteristics of each distribution. Fourth, it will be necessary to measure and analyze selected variables which affect the systems under study. Fifth, several systems will be examined in a diachronic perspective to understand systemic stability.

In the sections of this chapter that follow, the space utility function of exchange and how it is manifested in the archaeological record are discussed. Then the context of studying exchange in California is examined to provide a background to the analytical data.

In chapter 5, existing methodologies for describing exchange are discussed in terms of their descriptive and analytical capabilities. A new methodology is formulated to undertake the study of regional exchange. Synagraphic mapping of a sample of archaeological data is introduced as a technique to describe the percent of obsidian in the chipped stone tool assemblages in the Late Horizon. The regional patterns define the exchange and use of obsidian from sources in California. These patterns are then compared to several other regional patterns representing variables linked to space utility. First, the major trails are compared to the trends of the regional exchange patterns. If the trails are related to regional resource distribution, they should link major resources, or "sources," to areas where these items are necessary, or "sinks." Also, the trails should act as lines along which resources are most readily distributed as exchange items. Second, the lithic resource base is compared to the patterning of the exchange systems. If the exchange of obsidian serves an important space-utility function, then one would expect that the patterning of use and exchange of obsidian would be greatly affected by the presence or absence of alternative lithic resources. Third, the presence of ethnolinguistic group territorial boundaries are examined and compared with anomalies or changes with the regional exchange systems. If social interaction is more important than space-utility function of exchange, then one should observe the effect of boundaries on social interaction. In all, chapter 5 introduces a new methodology for describing regional systems, and then examines three parameters of the systems to determine their importance as mechanisms of the space-utility function of exchange.

In chapter 6, analysis shifts from a broad regional perspective to that of source-specific exchange systems. The original data are further refined by using chemical characterization data. Chapter 6 demonstrates the methodology designed to isolate individual exchange systems. Then each system is examined to provide a comparative systems analysis. Three variables considered to be manifestations of space utility are quantitatively analyzed to determine their respective importance.

In chapter 7, the quarry production rates of three exchange systems are examined in order to understand how the systems interact with each other. Obsidian hydration dating plays an important role in ascertaining the chronological framework of this study. Quarry production analysis, developed by Singer and Ericson (1977), is used to describe the production of obsidian preforms for each of the three obsidian sources. These data indicate the degree of interaction between the sources as well as elucidate some of the unresolved questions which have perplexed students of California prehistory for many years.

Chapter 8 is the summary-conclusion of this monograph; it outlines the findings involving obsidian hydration dating and the importance of exchange as a space-utility mechanism.

THE SPACE-UTILITY FUNCTION OF EXCHANGE

It is necessary to define what is meant by the space-utility function of exchange. Originally, both time and space utility were concepts proposed by Binford (1965) as mechanisms used to conserve energy and matter: "Space utility is gained when energy and matter can be put to work over a greater geographical area by transporting them beyond the geographical area from which procured" (Binford 1967).

This definition of space utility is very limited in the explanatory sense since a simple demonstration of the existence of nonlocal materials would satisfy the definition. What seems to be missing is a suggestion of why space utility should occur and the degree of space utility gained. Basically space utility is gained through extending the resource base of the region, but the degree of space utility gained depends upon such things as conformance to Zipf's (1949) Princple of Least Effort, or more recently, the minimaximization of energy (Von Neumann and Morgenstein 1955). Since in most regions raw materials are unevenly distributed across the landscape, the effective gaining of space utility is necessary to acquire the needed resources. It is hypothesized that an efficient method is developed to accomplish this acquisition process. Some of the possibilities are the following: the development of regional exchange systems, the use of existing trails as transportation routes, the use of the nearest source of a given material or similar substitute material, and the effective minimaximization of distance between sources of different material, and so on. In the following pages the efficiency of each process or method will be examined.

First, consider whether the development of a regional exchange system is a relatively efficient means to acquire regional resources. For any given class of raw materials there are usually a number of alternatives, so that the selection of a particular material must be the result of evaluating a series of contingencies. Three contingencies considered here are the relative efficiency of the material in performing specified tasks (Ericson and Singer 1977), extraction cost, and "work expenditure." Work expenditure involves both horizontal and vertical distance traversed in transportation of the given load (Ericson and Goldstein 1980).

To each point in space, and for each material of each class, a particular contingency value can be assigned. It remains to be shown whether these values are determinants in the procurement of specific materials at certain points. If so, they also may have served an important role in the formation of a regional exchange system. Thus, it is expected that the distribution of resources of a region, i.e., the location of raw materials necessary for subsistence or technology, will have primary importance in determining the internal organization of a regional exchange system. The efficiency of such a system can be compared to alternative means of acquiring regional resources. For these purposes it is important to consider the modes of acquisition: (1) acquisition of a material through direct access by the population, and (2) acquisition of the material through a regional exchange system. In this discussion, the former will be called the <u>direct access model</u> and the latter the <u>exchange model</u>.

The direct access model describes the direct acquisition of a resource by local people, or common resource acquisition. The consumption, distribution, and production of resources reside with the individual or local group. The behavioral basis for the operation of this model is simply illustrated: "An individual who needs the resource for himself or his local group simply walks to the nearest source, over a particular route." Most likely the route is chosen to minimize "work." The probability of even going to the source and undertaking this action is assumed to be a function of two variables: (1) "work expenditure" of undertaking the action, with distance as an estimator of work expenditure, and (2) the relative importance of the material to the individual or local group. Given a homogeneous plane, and a series of local groups having similar technology within the region, the simulation of the direct access model would show the decrease in the amount of the resource in space as a series of concentric circles representing decreased amounts of the material as some function of "work expenditure" weighted by the importance of the material to the local group. Perhaps this is similar to the spatial demand cone around a retail store shown by Berry (1967) to follow exponential decay. But Renfrew (1977) does provide a number of alternative mathematical models describing particular modes of behavior.

The exchange system model describes the acquisition of a resource through trade partners within the system. The modes of consumption and distribution involve a system of trade partners, who, in the exchange process, remove a portion of the material they receive. The behavioral basis for the operation of this model is simply illustrated:

> An individual who needs the resource for himself or his local group will obtain it through exchange with his trade partner. The repetition of this event throughout the system promotes the movement of the item through the system.

Both models imply a gain in space utility. However, there may be a quantitative difference in the efficiency or energy requirements between the modes of acquisition. It is expected that a regional exchange system is more efficient than acquisition by direct access. "Work expenditure" or transport costs which have been specified as an important variable for the direct access model can be used as a measure to compare the efficiency of the two models. Suppose there are two resources, obsidian and salt. Moreover, there are seven or n villages, each separated linearly by a unit distance, X. The end villages are the sources of obsidian and salt, respectively. This arrangement is illustrated in plate 4-1. If obsidian and salt are acquired through direct access by each village separately, then at least m individuals, one from each village, must at some time travel from the village to salt to obsidian. A crucial fact is that the trips _to_ each source are made without a load. On the other hand, if obsidian and salt are acquired through trade partnerships within an exchange system, an individual travels to the next village and returns each time loaded with the opposite item. The total regional transportation cost per unit time per unit load is up to twice more for direct access than through exchange. While the factor of two is a crude comparison of the two models, it does illustrate the fact that there is increased energy cost of acquiring items through direct access rather than through exchange which would favor the development of a regional system.

Second, it is important to examine if indeed existing trails are used as transportation routes. The efficiency of each trail can be measured using a least work model considering both vertical and horizontal work space. Such comparative evaluation is beyond the scope of this study. However, if the trails are related to regional resource integration, they should link up major resources. The major trails and regional trails of the regional exchange patterns are examined in chapter 5.

Third, the use of the nearest source for a given material (or substitute material) implies that a choice is made between materials of the same class or similar classes. The use of the nearest source of a given material (or substitute material) is considered to be efficient in conserving energy consumed in transport. On a regional level, space utility is gained, and in the physical sense, energy is ultimately conserved. It is expected that interactions will be shown between materials of the same class (e.g. obsidian) and their substitutes (e.g. chert). This aspect is examined in chapter 5. Likewise, it is expected that between materials of the same class there is also a certain degree of interaction. This form of interaction is examined in chapter 6. Finally, the effect of the "next nearest" source is examined as a variable in chapter 6.

Fourth, the influence of distance from a source as affecting amount of a given item obtained is another aspect of gained space utility while at the same time conserving energy. These relationships will be examined in both chapter 5 and chapter 6.

Fifth, the effective minimaximization of distance between sources of different material is an important manifestation of gaining space utility; however, its testing is beyond the scope of this study at this time.

In conclusion, the space-utility framework within which to examine the internal structure of regional systems or methods used to acquire regional resources opens up many new areas for research. Here, space utility is more than just an expansion of a resource base; it also involves the ways in which energy is conserved in that expansion. The succeeding chapters will examine in detail specific aspects and analyze particular variables relevant to gaining space utility.

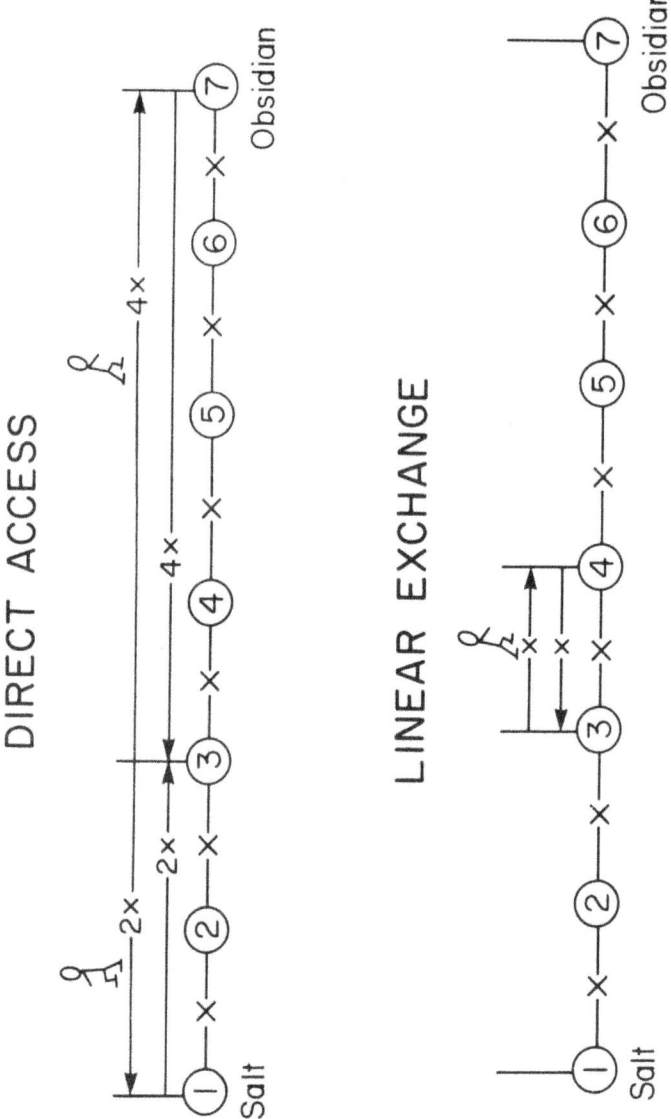

Plate 4-1. Systemic Work Expenditure in Resource Acquisition by Direct Access and Linear Exchange. This diagram presents a simplified illustration of the differences in unit travel cost (work expenditure) involved in acquiring two different resources.

THE CONTEXT OF STUDYING EXCHANGE IN CALIFORNIA

California can be seen as a unique region within which to study the phenomena of exchange systems on a continuing basis. Its environmental diversity and rich anthropological records offer an ideal setting in which to evaluate the conditions and factors which favor the development, maintenance, and stability of exchange systems that have been documented for this region (Davis 1961). The high diversity of the environment, lithic sources, and biological communities provide sufficient variation to permit study of the effects of different variables on the operation of exchange systems. Future analysis will determine the degrees of importance of the space-utility function of exchange systems. Complementary to these investigations, ample ethnographic, ethnohistoric, and archaeological data are available on the aboriginal, hunter and gatherer populations. Although the complex forms of regionally centralized organizations, such as the redistributive chiefdoms of Hawaii (cf. Earle 1977a) or the Northwest Indian groups, are not present here (compare Frederickson 1971; C. King 1971, 1973; T. King 1970, 1972), there appears to be evidence for a complex integration of people and resources beginning at least by 500 B.C. (Frederickson 1973b), if not during Windmiller culture prior to 2000 B.C. (Singer and Ericson 1977; Heizer 1974). Indeed, without centralized organization, these low-ordered systems were integrated by exchange between trade partners and independent sociopolitical units (cf. Kroeber 1925; Davis 1961).

The tremendous diversity in California landscape, resources, and environment offers a good opportunity to observe the effects of different variables in the operation of exchange systems. (For instance, the topographical variation of California allows an evaluation of Zipf's [1949] principle of least effort as it applies to the location of trails through varied terrain.) An example of the magnitude of variation is demonstrated by the Central California regional along west-to-east transact. The topography ranges from sea level in the San Joaquin Valley to the 11,000 feet passes through the Sierra Nevadas to the 7,000-9,000 feet of the Bishop plateau down to 2,000-3,000 feet in the valleys of the Basin and Range Province. Within the boundaries of California, most of the ecotones of the temperate zone (cf. Shinem 1972) are represented, providing sufficient biotic and environmental diversity for studying the effect of resource diversity and broad interaction between biotic resource zones. One might expect that the intensity of exchange would be greater crossing perpendicularly to such zones than parallel to them. As a result, an exchange system would be asymmetrical, e.g., showing more interaction in a direction perpendicular to ecotone boundaries. Ecologically, California can be defined in a general manner as a series of parallel North-South biomes conforming to the topography. These offer extreme diversity of biotic resources: marine, coast, chaparral, grassland, marsh, coniferous forest, alpine, high and low desert (Odum and Odum 1959:387). The mineral resources of California are equally diverse. The distribution of aboriginal utilization of mineral resources has been established (Heizer and Treganza 1944) and, more recently, many of the mineral sources have been chemically characterized for exchange system analysis (Jack and Carmichael 1969; Jack 1976; Jackson 1974).

The early travelers' accounts and the extensive ethnographic descriptions by Kroeber (1925), Steward (1933, 1938) and many others provide the data particularly useful for this research, such as the ethnographically-recorded territories of ethnolinguistic groups (Heizer 1960; Bennyhoff 1977) and their respective population size (Kroeber 1925; Cook 1943, 1955a, 1955b; Heizer 1960; Baumhoff 1963; Brown 1967; Bennyhoff 1977). It is important to note that by historic times not less than 10% of the total Indian population lived in California which comprises approximately 1% of the total area of North America north of Mexico (Heizer 1964:120). Here data permit evaluation of the effects of territorial boundaries and population density on trade systems. Additional data for studying exchange are provided in a compilation of the items exchanged, the directionality of group reciprocity, and location of aboriginal trails (Davis 1961). Research originally concentrated on shell trade with the Southwest (Brand 1938; Colton 1941; Heizer 1941a; Tower 1945). Research has recently expanded to include general studies on exchange (Davis 1961; Vayda 1967; Price 1967; Frederickson 1969; Chagnon 1970; Bean 1971; Bettinger and King 1971; Ericson 1973b; Jackson 1974; Jack 1976).

Archaeological information for California is abundant. Millions of man-hours have been invested in site survey and excavation; extensive archaeological collections of records are available, numbering at least 500 site reports and manuscripts. Here data permit exchange system analysis.

Unfortunately, despite all of the man-hours involved, there has yet to appear a regional synthesis which would serve to unify archaeological investigation in California (Heizer 1964:136; Warren 1973:231). This may be due in part to the complexity of the ethnohistorical record which is the result of the considerable linguistic, cultural, and material variability on both the ethnographic and the prehistoric time levels of this West Coast region (Heizer 1964:118). An acceptance of the "fish-trap theory" helps account for this observed diversity where the "multiplicity of langauges in California is due to the successive crowding, into this more desirable habitat, of waves or bands of unrelated immigrants from less favorable territories, to which none of them were ever willing to return" (Kroeber 1962). In part, the explanation is methodological. As indicated by the history of archaeological investigation in California (Warren 1973), there has been a concern with local sequences for the cross dating of archaeological sites and sectional space-time systematics with the development of the cultural horizon system supported by relatively few radiocarbon dates (cf. Ericson and Hagan n.d.; also compare appendix 2).

Local sequences as noted by Warren (1973) have been established for Central California (Lillard, Heizer, and Fenenga 1939; Beardsley 1948, 1954; Bennyhoff and Heizer 1958; Cook and Heizer 1962; Ragir 1972; Frederickson 1973a), the North Coast Range (Harrington 1948; Heizer 1953; Meighan 1955c; Meighan and Haynes 1970; Frederickson 1973a), Northeastern California (Fenenga and Riddell 1949; Riddell 1956, 1960; Baumhoff and Olmsted 1963; O'Connell 1968; O'Connell and Ambro 1968), the Sierras (Heizer and Elsasser 1953; Elsasser 1960; Bennyhoff 1958; Fitzwater 1962, 1968), Owens Valley (Lanning 1963; Michels 1965; Clewlow, Heizer, and Berger, 1970), California desert (Rogers 1939, 1945; Harner 1958; Hunt

1960; Wallace 1962; Bettinger and Taylor 1974), and South Coast (Rogers 1929; Wallace 1955; Warren 1968; King, Blackburn, and Chandonet 1968; King 1971).

A brief and oversimplified version of the cultural history for Central California from the Tehachapi Mountains to the head of the Sacramento Valley, and the Pacific to the Sierra Nevada is paraphrased here (Heizer 1964). The archaeological record of Central California can be divided into three broad geographical zones: the coast, the interior valley, and the western foothills of the Sierra which are important for local adaptation. The Early Horizon is typified by charmstones, extended burials, large projectile points, use of the atlatl (Heizer 1949), flat slab metate and bowl mortar, and several types of shell beads. Individual wealth accumulation did not exist and most likely warfare was uncommon. The Middle Horizon is an outgrowth of the Early Horizon culture from about 2000 B.C. to 300 A.D. Some of the observed changes are tightly flexed burials with minimal offerings, rare cremations although usually rich in offerings, a large variety of abalone-shell beads, deep wooden mortars, barbed harpoons and blunt-tipped bone or antler point, probably for taking fish (Bennyhoff 1950). The bow appears along with the atlatl which was apparently used for special purposes in hunting or war. Human bones are fairly frequently found with imbedded points, indicating existence of warfare. In general, the sites are larger and the population was greater than in the Early period. The Late Horizon, beginning at about 300 A.D. and continuing until the late eighteenth century, can be identified with prehistoric culture of the Penutian-speaking tribes of Central California (Heizer 1941b). Although the Middle Horizon material culture overlaps with that of the Late Horizon, there are specific changes registered in the archaeological record. Cremations with burned grave goods increase. Large stone mortars, long steatite smoking pipes, incised bird-bone tubes, and an array of new shell beads and abalone ornaments make their appearance. Of importance here, small obsidian arrowpoints (often with deep edge serration) were used with bow and arrow. During this period (about 1300 A.D. in the southern Sierra area) Central California comes under the influence of the "Greater Southwest," based on both ethnographic and archaeological evidence (Heizer 1946; Haekel 1958; Taylor 1961; Klimek 1935; Kroeber 1923). This line of demarcation, the northwestern frontier of the "Greater Southwest," is coincident with the territories of the Pomo, Wintun, and Maidu tribes (Heizer 1964). As will be shown, the exchange systems which existed most likely served an important function in the transmission of materials and ideas represented in the archaeological record of this area. Heizer (1964) suggests a general picture of the coast and valley of Central California as a change from peaceful egalitarianism in the Early Horizon times to warlike wealth-consciousness in the Middle Horizon. The Late Horizon appears to be conformable to that recorded by ethnographers like Kroeber (1925, 1962). Population begins to increase in the Middle Horizon and continues with improved subsistence techniques until the historic period when perhaps 10% of the total North American Indian population inhabited approximately 1% of the area. It is noteworthy that this phenomenal population density existed among hunters and gatherers without agriculture.

A review of the historic period is made since the presence of ethnolinguistic group territorial boundaries are examined and compared with anomalies or changes with the regional exchange systems. If social interaction is more important than the space-utility function of exchange, then one should observe the effect of boundaries on social interaction. Kroeber (1925:899-918) recognized four types or provinces of ethnic culture which are paraphrased here.

The Northwestern California culture was the limited area occupied by the Yurok, Karok, and Hupa, which extended northward into Oregon. These groups were related to the culture of the North Pacific coast which included other groups like the Kwakiutl and Haida. In California, the Klamath River, with its rich salmon population, acted as an important resource base for these people. The people of this area were mainly Athabascan speakers with the Algonkins culturally dominant. Adjoining to the east, Kroeber describes a Lutuamian or Klamath Lakes culture or subculture represented by the Klamath-Modoc, Shasta, and Achomawi.

The extent of the Central California culture province conforms to that described above. If there had been a definable cultural center of this ethnic province, it would lie between the Pomo, Patwin, and Valley Maidu. Whereas the northern and southern ethnic provinces are considered marginal areas or extensions of greater cultural regions, Central California culture is considered to be an isolate. Kroeber does include part of the Great Basin as a subcultural area with close cultural kinship. Central California was distinctly a Penutian language center with Hokan fringes. The resources of the Sacramento and San Joaquin Rivers and their tributaries formed an important resource base for this area.

The Southern California cultural province centers near the coast including the Chumash, Gabrieliño, and Luiseño. The dominant language group was Shoshonean with Chumash (Hokan) represented on the coast. Both maritime interior resources were exploited aboriginally.

The Lower Colorado River cultural province was an extension of the great Southwestern cultural area, which included the Mohave, Yuma, Chemehuevi, and Diegueño. The dominant language is Yuman with Shoshonean spoken on the margins of the ethnic province. Agriculture was practiced here on the bottomlands of the Colorado which overflowed annually. Kroeber suggests that both the Southern California and Lower Colorado cultures are genetically related to the Pueblos or their ancestors. Thus, California provides an excellent region for studying exchange given the tremendous diversity of cultures, environments, and resources, which were interlinked by exchange.

The ethnographic record of aboriginal California does not provide the data necessary for the reconstruction of regional exchange systems. These data do document that exchange existed between particular groups and the items that were exchanged. Very little information is available on the amounts exchanged or the spatial or temporal context of the exchange events. Davis (1961) has assembled import-export item inventories for each ethnolinguistic group based on ethnohistorical records. Table 4-1 shows that food, shell money, and obsidian are the three major types of exchange items. These records of transactions and material transfer are discontinuous between groups so that it is impossible to predict where a given item is introduced

Table 4-1

NUMBER OF GROUPS REPORTING SPECIFIC EXCHANGE ITEMS
MENTIONED IN THE ETHNOGRAPHIC LITERATURE AS
BEING IMPORTED AND EXPORTED
(after Davis 1961)

	Item	Imported	Exported
1.	Salt	39	40
2.	Basketry	40	36
3.	Hides and Pelts	38	30
4.	Marine shell beads (other than listed)	30	28
5.	Acorns	24	23
6.	Dentalia	23	22
7.	Clam disk beads	22	22
8.	Whole or fragmentary marine shells	21	21
9.	Fish	20	17
10.	Obsidian	19	16

into the system, how much is removed at each juncture, and the full extent of its distribution. This is an inherent weakness of the early ethnohistorical records. Nevertheless, several major trade routes have been suggested by the above data and are illustrated in plate 4-2. It is clear that some real information on the development and operation of the regional exchange systems is critical at this point using a different frame of reference.

Chapter 5

THE REGIONAL EXCHANGE OF OBSIDIAN IN THE
LATE HORIZON IN CALIFORNIA

INTRODUCTION

The value of a methodology which can describe the detailed spatial patterning of exchanged items within a given region is immense because it would permit the analysis of systemic variables responsible for the basic trends and anomalies observed. More importantly, however, this analysis would permit the isolation and ranking of these systemic variables as to their effect on exchange. Such a methodological framework would also be amenable to the study of a regional exchange system and its development in an evolutionary framework which incorporates analysis of space-time, systemic, and cultural variability. Toward these desired ends, the exchange systems in California, defined by the prehistoric utilization of a given obsidian source, are described and analyzed, using the methodological framework developed in this preliminary study.

The quantitative analysis of prehistoric exchange systems is a recent and developing field in archaeology. Beginning with the work of Renfrew, Dixon, and Cann (1968), researchers have evaluated exchange systems by the observed changes in the quantity of an exchanged item as a function of distance from its source. In this method, an exchange index, representing the quantity of a particular item in its archaeological context, is plotted or statistically regressed as a function of distance between the point of observation and its specific source. These analyses of distance-dependent trends in regional exchange systems have been useful for comparative analysis. For example, Hodder (1974) has investigated differences in transport systems reflecting the bulk weight of the exchanged items, and Sidrys (1977) has identified the hierarchical ranking of central places within a regional trend. Although this methodology, termed two-dimensional analysis, is frequently employed in quantitative studies, it has several limitations.

Since in two-dimensional analysis only the magnitude of an observation and its distance from a source is considered, the spatial position of the observation is not considered in its local context, so that this simplification often masks significant variability in the data. For example, although a regression line can be fit to the data, the statistical "outliers" and other data may represent other significant trends which can be easily overlooked by the analyst. The analyst should expect that within a spatial data array, each observation represents orders of variability: (1) "large-scale" systematic changes, such as the decrease of an item as a function of distance or the Law of Monotonic Decrement (Renfrew 1977); (2) "small-scale" fluctuations, including (a) nonsystematic variation, such as changes in the quantity of an item due to effects of a central place hierarchy (Sidrys 1977) and (b) systematic local variations, such as changes in the quantity resulting from the bifurcation of a network system; and (3) chance variation, such as would result from sampling error (Krumbein and Graybill 1965). The collapsing of orders of variability with the resultant masking of other important trends in the data imposes a definite limitation on two-dimensional analysis.

The validity of two-dimensional analysis is dependent on the overall symmetry of the regional system, i.e., that there is more or less an equivalent decrease in the quantity of the item as a function of increasing distance, regardless of the direction from the source. For symmetrical systems, linear regression analysis can be used to define the regional trend. However, in asymmetrical or directed systems one would expect that linear regression of distance would be inappropriate. For example, in the analysis of an asymmetrical system, many regional trends could be defined, depending on the selected sample. This raises the question—Which trend would be most characteristic of the system or "type of system" under investigation? It is maintained that all the directed trends are significant in terms of understanding the operation of multiple systemic variables. In this perspective, the operation of these yet unidentified parameters may be the direct variables of a system, whereas, in such cases, "distance" may be an indirect variable. For example, population and its fluctuation is shown as an important variable in utilitarian exchange (cf. Wright and Zeder 1977).

In summary, the masking of significant sources of variability in the data, the dependency of the results on the symmetry of the system, and the significance of the distance-dependent trends impose major limitations on the applicability of two-dimensional analysis. The need for the development of a different methodological approach is apparent, particularly one which overcomes these limitations.

Unlike two-dimensional analysis, the new method is characterized as a three-dimensional approach. This process describes the spatial patternings of the exchange items, the degree of symmetry of the exchange system, the regional trends, local trends, spatial anomalies, and to some degree the nature of the observational error. With this comprehensive description of the system, both qualitative and quantitative analysis of the variables operating on the system can be conducted.

The operation of a set of independent variables can be qualitatively defined by inspection or by superimposition of data over the spatial patterning, with particular focus on the observed trends and anomalies. The final phase, the testing of these variables, can be accomplished by multiple regression analysis.

The methodology proposed in this chapter is applied to archaeological and ethnohistoric data in California. The details and results of its application are discussed in the following sections.

A DESCRIPTION OF EXCHANGE SYSTEMS

In three-dimensional analysis, the need for a technique which describes the spatial distributions of exchanged items within their local context has been established. The technique utilized in this section is borrowed from locational geography. Specifically, synagraphic mapping by the use of the computer program, SYMAP, developed by Fisher (1973) and modified by Lankford (1974), has been designed to display, relate, and weigh spatial data. Through interpolation, SYMAP describes the distributions of an exchanged item in space. SYMAP performs the interpolations of the data through multiple regression analysis. In general, the interpolation of the value of a given point is determined by the values of seven nearby data points. The interpolated values are assigned to a contour map. Thus, synagraphic mapping of exchange data is the vital first phase of three-dimensional analysis.

Four basic steps in sampling and compilation of the exchange data are used to produce the exchange distribution map (plate 5-1). The initial step is to select an exchange item for analysis. Obsidian was chosen because many of its characteristics are ideal for a regional study: (1) the high chemical and physical durability of obsidian (cf. Ericson et al. 1975) contributes to its preservation in archaeological sites; (2) its use as a raw material in the manufacture of chipped-stone tools appears to be utilitarian and, as such, can be compared to the use of alternative lithic materials for internal consistency; (3) obsidian and other lithics were collected with relatively little bias by the archaeologist; (4) there are many obsidian sources, well distributed throughout California (cf. Ericson, Hagan, and Chesterman 1976), providing an opportunity to study interaction among sources; (5) the particular source of obsidian can be identified by both X-ray fluorescence (Jack and Carmichael 1969; Jackson 1974; Jack 1976) and instrumental neutron activation analyses; and (6) there exists a potential for direct dating by obsidian hydration (cf. Ericson 1975).

The second step is to establish an exchange index. Several different types of indexes can be used as measures of exchange, each having certain advantages and disadvantages. The exchange index, as defined by Renfrew, Dixon, and Cann (1968:327) is the percentage of obsidian in the total chipped-stone-tool category at each site. This index was selected as having the least bias introduced by techniques of recovery. It is important to note that the index represents the relative degree of the occurrence of obsidian rather than its absolute quantity.

The next step is to select and tabulate data for synagraphic mapping. Fifty-two Late Horizon sites, tabulated in table 5-1, were chosen to give an area coverage using the compiled site information file (Ericson and Hagan n.d.). The following information on each site was recorded: the percentage of obsidian tools in the total chipped-stone-tool category was calculated as the exchange index from the raw material specifications; and the sites were located and assigned coordinate numbers within an X-Y rectangular grid system using the base-meridian intersections and the plane coordinate intersections (U.S. Department of Commerce 1954) to minimize curvilinear distortion.

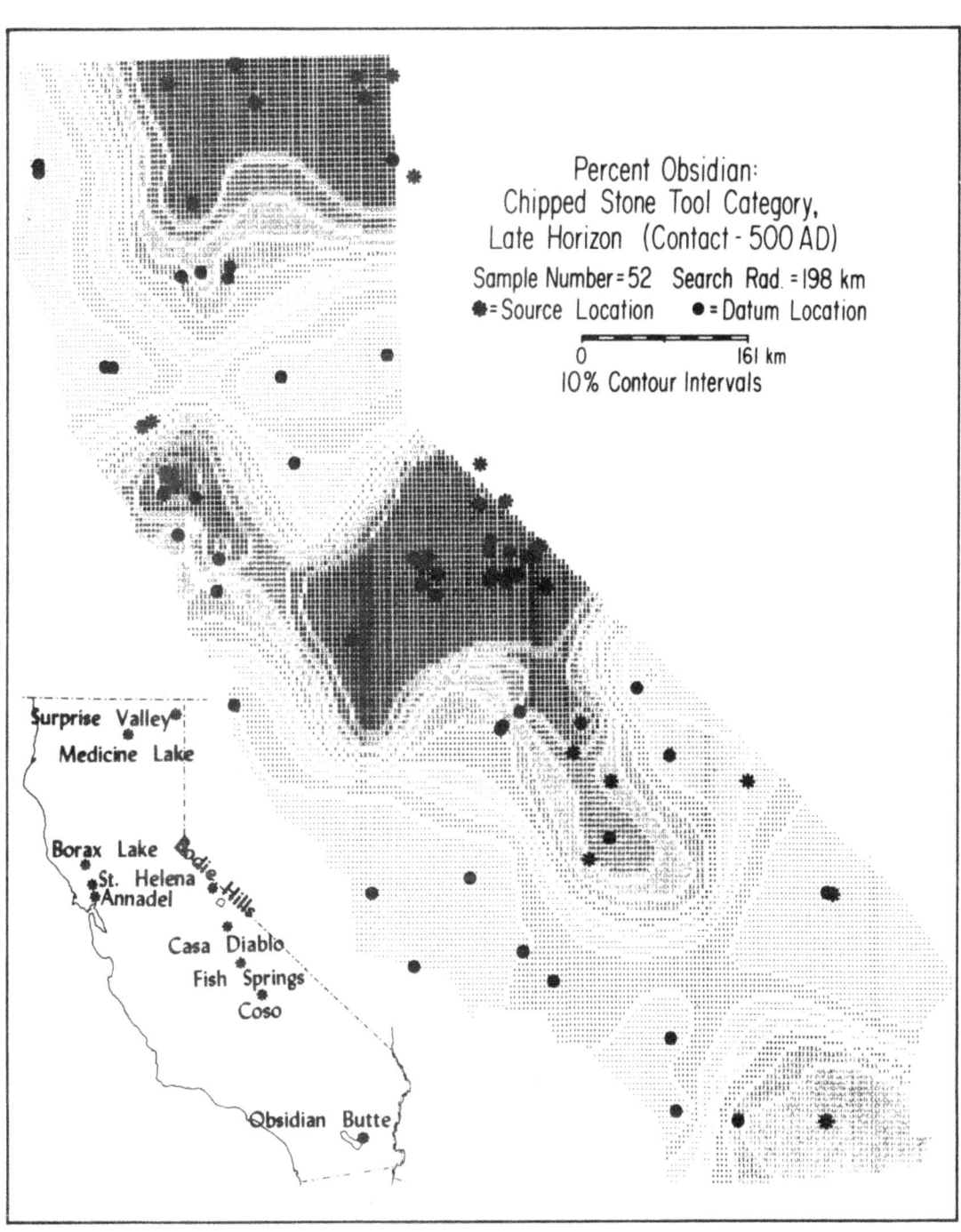

Plate 5-1. A synagraphic map of the prehistoric egalitarian exchange systems in the Late Horizon of California. The 10 sources enumerated in the bottom right-hand corner are evaluated through systems analysis.

Table 5-1

REFERENCE INDEX TO DATA USED IN SYMAP

Site	Reference
Mrp-56	Rasson 1966
SBa-54	Harrison and Harrison 1966
Ballerat*	True et al. 1967
LAn-167	Ruby 1966
Hum-118	Elsasser and Heizer 1966
Hum-169	Elsasser and Heizer 1966
Tul-1	Pendergast and Meighan 1959
Mrp-181	Fitzwater and Van Vlissengen 1960
Las-194	O'Connell and Ambro 1968
SDi-474	Townsend 1960
SDi-473	Townsend 1960
SDi-655	Wallace 1960
Ala-328	Davis and Treganza 1959
SBr-288	Davis 1962
Nev-15	Elsasser 1960
Sie-20	Elsasser 1960
Iny-2	Riddell 1951
Teh-193	Baumhoff 1957
Men-500	Meighan 1955c
Teh-1	Baumhoff 1955
Iny-222	Meighan 1953b
Nap-1	Heizer 1953
Nap-14	Heizer 1953
Sha-20	Smith and Weymouth 1952
Ker-40	Wedel 1941
Sha-47	Smith and Weymouth 1952
Sis-13	Wallace and Taylor 1952
Mrn-115	Meighan 1953a
Childago Canyon*	Meighan 1955a
Benton Range*	Meighan 1953a
Crooked Meadow*	Meighan 1955a
E. Walker River*	Meighan 1955a
N. Owens River*	Meighan 1955a
Mrp-97	Bennyhoff 1956
Mrp-105	Bennyhoff 1956
Teh-58	Treganza 1954
Gle-15	Treganza and Heicksen n.d.
English Ridge Res.*	Childress and Chartkoff n.d.
Mnt-371	Evans n.d.
Sac-166	Gebhardt n.d.
Mno-382	Michels n.d.
SLO-297	Smith and LaFaue n.d.
SLO-298	Wire n.d.
Tul-24	Von Werlhof n.d.-2

(contd.)

Table 5-1 (contd.)

Site	Reference
Tul-145	Von Werlhof n.d.-1
Mrp-9	Bennyhoff 1956
Ker-62	Kowta n.d.
Tuo-236	Fitzwater n.d.-2
Sis-258	Johnson n.d.
Riv-463	Wilke n.d.
Fre-115	Lathrap and Shutler 1953
CCo-309**	Frederickson 1969
LAn-324	Ericson n.d.

*   Survey.
** Chipping waste.

examined by superimposing data on their distribution over the exchange system map.

Superimposition of the trails (plate 5-2), as described by Davis (1961), indicates that the gradients of several exchange systems coincided with the location of the trails. The notable examples are: (1) the north-south trail west of the Medicine Lake obsidian source, and (2) the two east-west trails over the Sierra Nevadas originating from the Casa Diablo obsidian source. These examples strongly suggest that the trails operated as lines for exchange. It appears that on a local scale the quantity of obsidian received is a function of distance from the trail and, possibly, the rank of the trail (primary, secondary, tertiary), as hypothesized from network analysis (cf. Irwin-Williams 1977). This evidence suggests that the best measure of distance would be a measure of the trail lengths.

For California, it was assumed that obsidian as a lithic material was used primarily in a utilitarian context. Thus, it was expected that the location of alternative lithic resources, such as other obsidian sources and alternative raw materials, would influence both the symmetry and extent of individual exchange systems. For these reasons, a selected portion of the regional geology of California was superimposed over the original exchange system map (plate 5-3). The Monterey and Franciscan formations containing local supplies of chert and the non-granitic portions of the Sierra Nevada batholith were selected, because they provided potential sources of alternative lithic materials for aboriginal California. It was impossible to determine potential sources of all lithic material for this report, but the three examples illustrate the effect that the presence or absence of alternative lithic sources had on the exchange of obsidian: (1) The most notable examples are the Trans-Sierran systems originating from Casa Diablo and Bodie Hills, where the granite of the Sierra Nevadas is the most common rock type. One might expect that the work expended crossing rugged terrain and addition limitations imposed by short snowless seasons would have discouraged the development of exchange over the mountains. However, the absence of alternative lithic materials appears to have been a dominant and overriding factor in the

At this point, the exchange index data and grid coordinates of the sites were entered as the data in the SYMAP program. In order to divide the range of the exchange index into equivalent values, from 0 to 100, a 10% contour interval was chosen to describe the spatial pattern of the distribution of obsidian. The resulting synagraphic map of the exchange index distributions is shown in plate 5-1. Two options of SYMAP were used to refine the map: (1) the outline of California was established by 84 points along its boundary, and (2) the coordinates of each obsidian source were entered as data to identify its location.

The synagraphic contour map (plate 5-1) describes the obsidian exchange systems of the Late Horizon in California. Some of the salient features of these systems are quite interesting.

The areas of highest percentage value of the exchange index enclosed the location of the obsidian sources. In other words, the obsidian sources appear to be the "sources" used in prehistoric obsidian exchange. If a major source of obsidian had been missed in prior obsidian source surveys, it would have "appeared" on the map.

The quantity of obsidian decreases as predicted by the Law of Monotonic Decrement (Renfrew 1977). However, the patterns are not symmetrical around the obsidian sources. In fact, the systems appear to be directed with regard to gradients defined as the direction of maximum rate of change. This "directedness" might be due to the locations of the sources of other exchange items, such as shell and salt, which have not been considered in this paper. Notably, the directedness is perpendicular to major ecozones.

There is no indication of the existence of hierarchical centers (or central places) which would be observed as abrupt, localized anomalies within the regional pattern (cf. Sidrys 1977). This result is expected since hierarchical-ranked central places are not considered to be characteristic of these exchange systems.

The size or extent of each system appears to be different. One would expect that the sizes should be equivalent or at least equivalent in filling the space between sources, assuming that the modes of production and transportation (through exchange) are approximately equivalent between systems.

The patterns suggest the overlap of many systems, as is particularly illustrated by the Trans-Sierran systems originating from Bodie Hills and Casa Diablo obsidian sources. In these cases, the discrimination of independent systems would require chemical characterization of artifacts. In sum, there are differences in the extent, shape, symmetry, and directedness which characterize each system. The variability of these properties suggests the operation of specific systemic variables within each system. In the following sections and in the next chapter, a number of variables are evaluated as to their effect on the systems enumerated in plate 5-1.

A QUALITATIVE ANALYSIS OF CERTAIN SYSTEMIC VARIABLES

As a preliminary procedure of analysis, certain variables which might have influenced the distribution of obsidian were evaluated qualitatively. The influences of trails, regional geology, and ethnolinguistic boundaries were examined by superimposing data on their distribution over the exchange system map.

Superimposition of the trails (plate 5-2), as described by Davis (1961), indicates that the gradients of several exchange systems coincided with the location of the trails. The notable examples are: (1) the north-south trail west of the Medicine Lake obsidian source, and (2) the two east-west trails over the Sierra Nevadas originating from the Casa Diablo obsidian source. These examples strongly suggest that the trails operated as lines for exchange. It appears that on a local scale the quantity of obsidian received is a function of distance from the trail and, possibly, the rank of the trail (primary, secondary, tertiary), as hypothesized from network analysis (cf. Irwin-Williams 1977). This evidence suggests that the best measure of distance would be a measure of the trail lengths.

For California, it was assumed that obsidian as a lithic material was used primarily in a utilitarian context. Thus, it was expected that the location of alternative lithic resources, such as other obsidian sources and alternative raw materials, would influence both the symmetry and extent of individual exchange systems. For these reasons, a selected portion of the regional geology of California was superimposed over the original exchange system map (plate 5-3). The Monterey and Franciscan formations containing local supplies of chert and the non-granitic portions of the Sierra Nevada batholith were selected, because they provided potential sources of alternative lithic materials for aboriginal California. It was impossible to determine potential sources of all lithic material for this report, but the three examples illustrate the effect that the presence or absence of alternative lithic sources had on the exchange of obsidian: (1) The most notable examples are the Trans-Sierran systems originating from Casa Diablo and Bodie Hills, where the granite of the Sierra Nevadas is the most common rock type. One might expect that the work expended crossing rugged terrain and additional limitations imposed by short snowless seasons would have discouraged the development of exchange over the mountains. However, the absence of alternative lithic materials appears to have been a dominant and overriding factor in the development of the systems. (2) It is also interesting to note the abrupt termination of these systems along the boundary of the Franciscan Formation, where extensive chert resources would have offered alternative materials. (3) The attenuation of the systems originating from the St. Helena, Annadel, and Borax Lake sources most likely can be explained by the presence of chert sources. When one considers the limited extent of these three systems, nearly at sea level (500-1,000 ft.), and the extent of the Trans-Sierran systems which begin at 8,000-10,000 ft., one is struck by the influence of alternative resources on the respective systems.

The important implication of these findings suggests that the resource base, i.e., the distribution of resources, is an extremely important factor in the development of utilitarian exchange systems which determines their extent and symmetry.

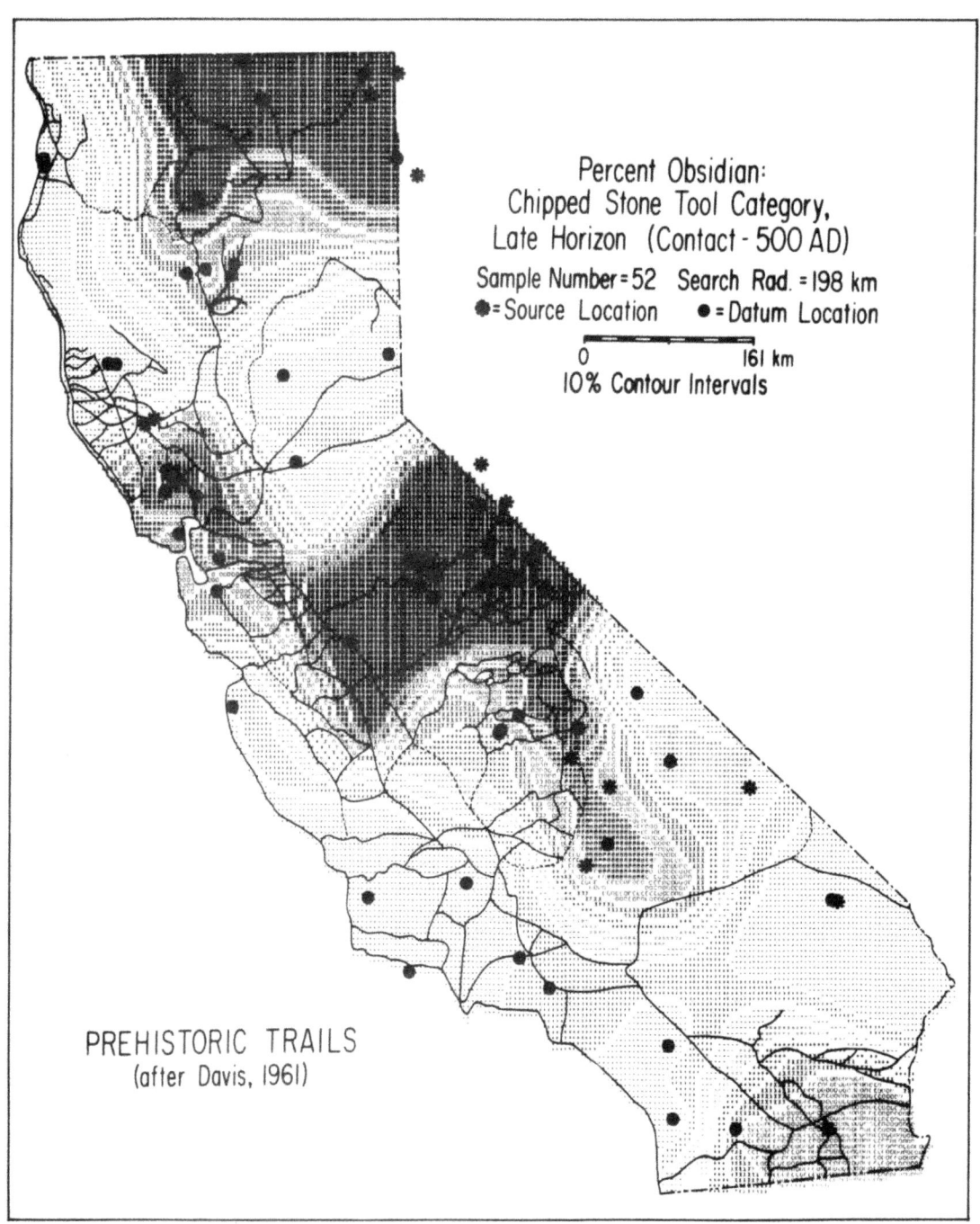

Plate 5-2. A demonstration of the correspondence of the major trails and gradients of the distributions of the obsidian within the exchange systems. The major trails are superimposed over the original synagraphic map. The trails were reconstructed notes and drawn by Davis (1961). Many of these trails conform to the location of modern roads.

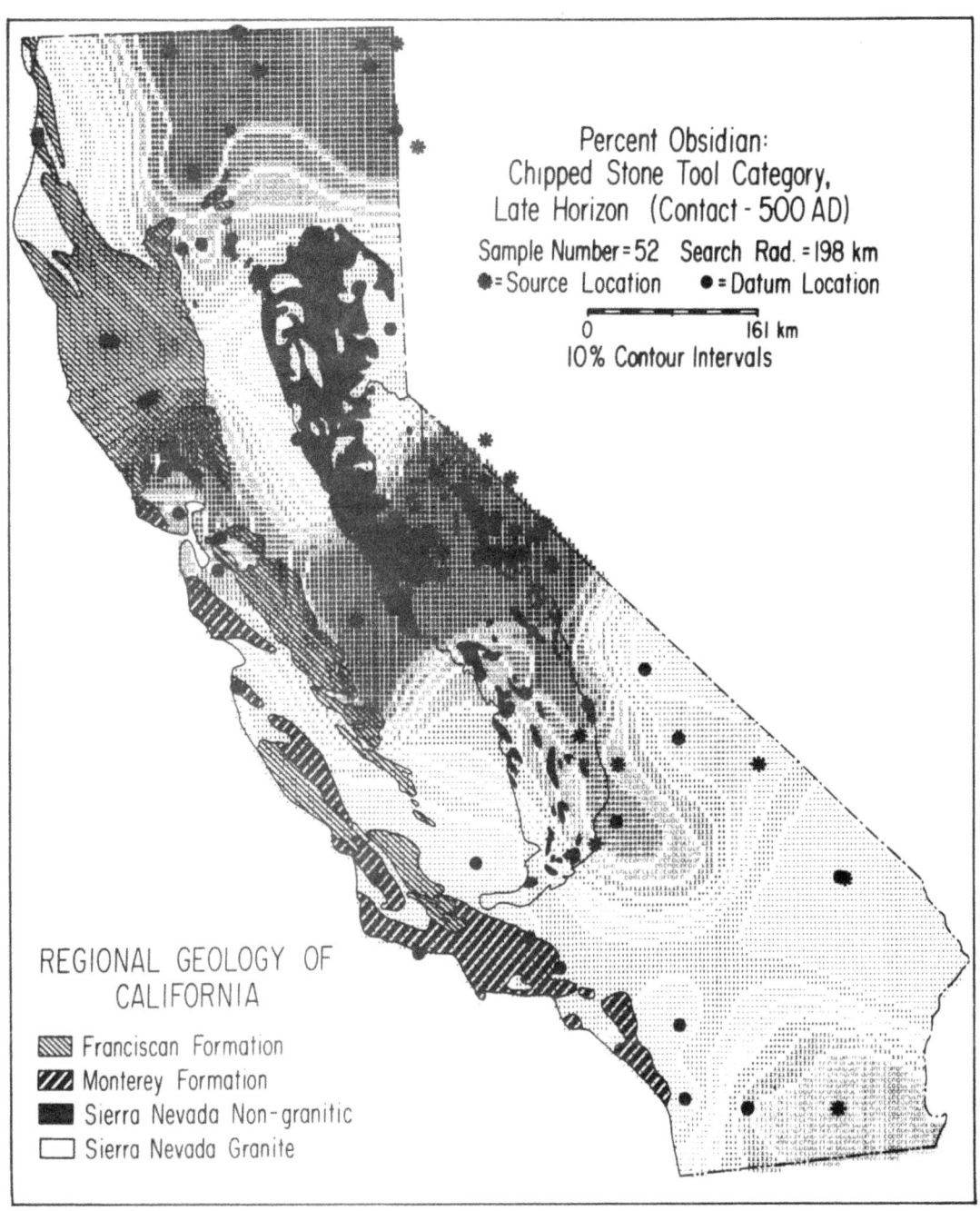

Plate 5-3. A demonstration of the effects of alternative regional resources on the distributions of obsidian within the exchange systems. The geological formation, containing abundant alternative materials for the manufacture of chipped stone tools, are superimposed over the original synagraphic map. The Coast Ranges along the western coastline contain Franciscan and Monterey cherts which were used for chipped stone tools. What is interesting here is the rapid decrease of the rise of obsidian as the Coast Ranges are approached. The Sierra Nevadas are composed of granite rocks which are almost unusable for chipped stone tools. Several areas within the Sierras have changes in the lithology, noted on the map, which again indicate an anomaly in the utilization of obsidian.

If it is assumed that the social distance of communities within an ethnolinguistic group is less than between ethnolinguistic groups, then it would be expected that discontinuities in quantity of exchange items would be observed at ethnolinguistic boundaries. For these reasons, the ethnolinguistic boundaries, described by Kroeber (1925:Map 1), were superimposed over the original exchange system map, shown in plate 5-4. The expected discontinuities do not appear. Although these results are discordant with Sahlin's model, in all fairness two additional factors should be considered. Kroeber (1925) describes that in many cases ethnolinguistic boundaries were mitigated by intermarriage between members of contiguous groups. Secondly, the original data set, selected to provide area coverage, does not necessarily provide sufficient control to resolve localized discontinuities. Nevertheless, the boundary effect does not appear to be important.

In conclusion, the distribution of resources is seen as an important factor in the development of exchange systems. The control of this variable will be important to gain a further understanding of the extent and symmetry of the systems and the process of exchange of utilitarian items. Secondly, the major trails appear to have served as lines of exchange. If this is correct, then the distance along the trails will be a better measure than straight-line distance currently employed in two-dimensional analysis. Thirdly, the existence of ethnolinguistic boundaries does not appear to affect the exchange of goods in the system. Finally, the qualitative analysis of the three selected variables demonstrates its value in evaluating and ranking variables prior to quantitative analysis.

CONCLUSIONS

It does appear that the proposed three-dimensional approach overcomes some of the basic limitations of former methodology. As applied to archaeological data of the Late Horizon of California, the technique provides a means to describe the spatial patterns of prehistoric exchange.

The results of the qualitative analysis of three selected variables are quite interesting. The presence and location of alternative lithic materials appear to determine the extent and symmetry of individual systems. Thus, the exchange of obsidian serves an important space-utility function, as it has been shown that the obsidian utilization is less in areas that have alternative lithic resources. Secondly, the correspondence between the major trails and gradients of the systems suggests that the trails served as lines for exchange. It does appear that the trails are utilized for regional resource integration. Also, the trails appear to act as lines along which resources are most readily distributed, which is another aspect of space utility. Thirdly, the observed lack of correspondence between the anomalies with the systems and the location of ethnolinguistic boundaries suggests that this observation should be reexamined with better data. If the exchange of lithic resources predominantly serves a space-utility function, then the above observations may be shown to be valid upon further analysis.

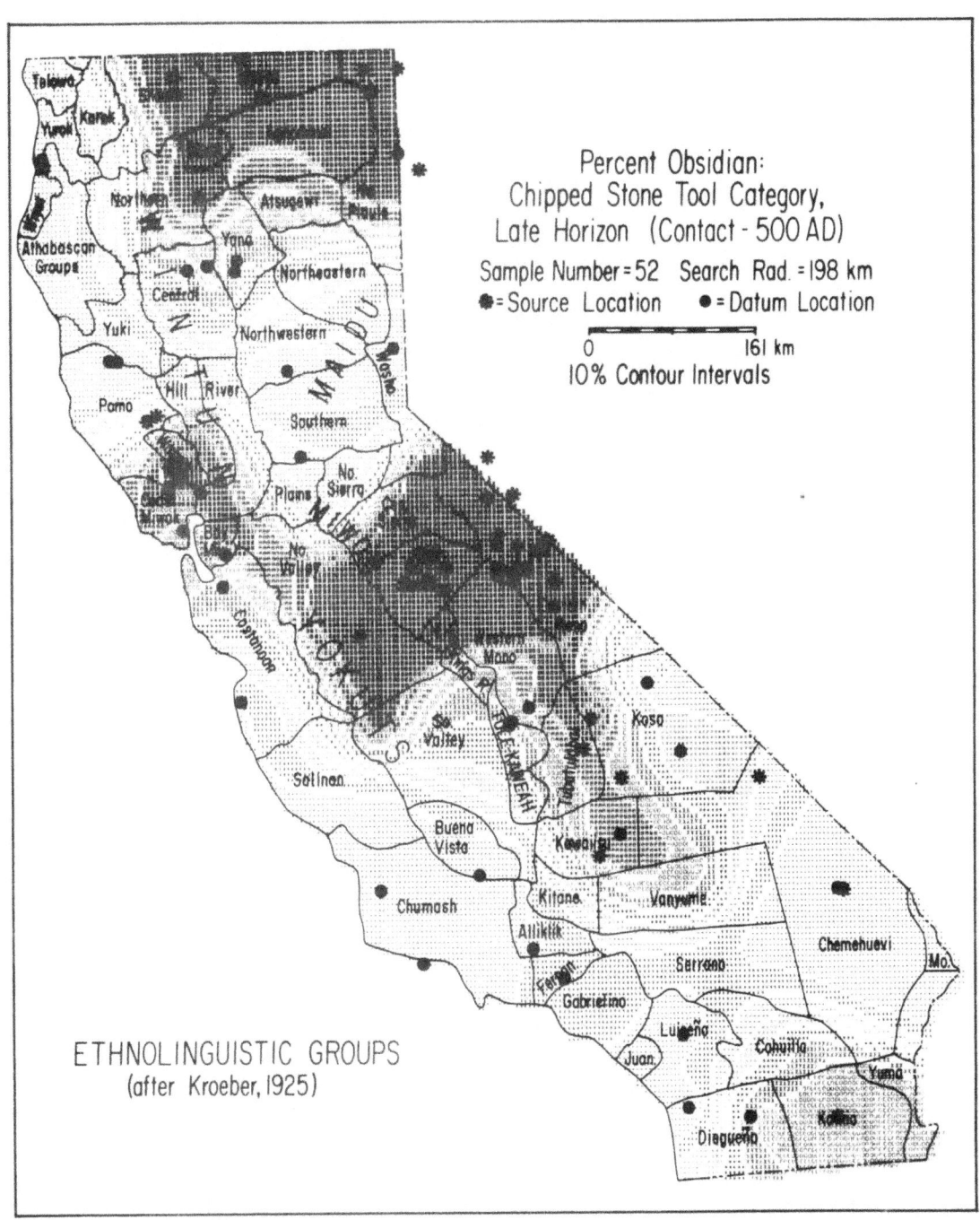

Plate 5-4. A demonstration of lack of correspondence between the boundaries of ethnolinguistic groups and changes in the distributions of obsidian within the exchange systems. The territories are superimposed over the original synagraphic map. The boundaries are drawn after those presented by Kroeber (1925). Although there is a lack of correspondence between changes in obsidian utilization and ethnographic boundaries using this regional data, this observation should be re-examined using well-controlled data sets.

In the next chapters, some other conditions and factors which favor the development, maintenance, and stability of exchange systems among hunter-gatherer economies will be examined.

Chapter 6

SOURCE-SPECIFIC EXCHANGE SYSTEMS:
A COMPARATIVE SYSTEMS ANALYSIS

INTRODUCTION

In chapter 4, a regional exchange system was defined as the distribution of obsidian, originating from a specific source, which is the consequence of the processes of exchange and utilization, and its subsequent deposition in the archaeological record. There is an inherent difficulty in calling the spatial distribution of one item a system. Such a definition runs the risk of generating hundreds of "systems" based upon the distribution of source-specific items. Nevertheless, for study purposes such distributions can be treated as systems in terms of their structure, organization, continuity, and geographical boundaries. The inherent difficulty with the presence of multiple obsidian sources is to identify the source of each artifact by some chemical process and then to specify the percent utilized at many locales. With the data thus controlled, source-specific exchange systems can be defined archaeologically. It is then possible to do a comparative systems analysis among the systems. Such an analysis will permit an evaluation of variables which control the morphology of a given system, such as its magnitude, shape, and geographical extent.

METHODOLOGICAL FRAMEWORK

In chapter 5, an exchange index map was made for the Late Horizon of California, the period extending from 500 A.D. to contact in the eighteenth century. There is a problem with the systems represented in plate 5-1, because with the present overlap of the systems, the exchange index cannot be evaluated for individual systems. Fortunately Jack (1976) identifies the percentage of each obsidian source utilized by each ethnolinguistic group in Northern and Central California, and these data allow the isolation of individual systems. Although this data set is not well-controlled temporally, in this preliminary analysis the effects of diachronic changes in utilization are presumed to be relatively unimportant because of the high percentage of Late Horizon sites in this sample.

A set of 10 source-specific exchange indices were calculated from the values of 121 arbitrarily selected data points on the original SYMAP shown in plate 5-1. These values were multiplied by Jack's data (Jack 1976) and the data sets were then entered into the SYMAP program, in order to generate the 10 source-specific exchange maps which are presented in plates 6-1 through 6-10. The percentage of a given contour interval is denoted as factor 1, 2, etc., instead of 10%, 20%, etc. The morphological properties of each exchange system and certain other archaeological details are presented in the next section.

NOTES AND DESCRIPTION OF EXCHANGE SYSTEMS

It is important to describe each exchange system in order to understand certain aspects covered in this dissertation. For these purposes, it is necessary to provide a detailed statement on how information for each characteristic of the systems was determined.

Under "Source data" information was derived from prior research (Ericson, Hagan, and Chesterman 1976) and field notes, presented in appendix 1. The "Source included" gives the name of the source or sources if more than one is included within the exchange system based upon Jack's (1976) chemical characterization scheme. The geographical location of the source is also presented (Ericson, Hagan, and Chesterman 1976). The remaining descriptive statements listed under "Source data" are derived from appendix 1 and data provided by Ericson, Hagan, and Chesterman (1976).

Under "System characteristics" the information was derived from inspection and analysis of the maps for each system presented in plates 6-1 through 6-10. "Rank size" is a number specifying the comparative rank from largest to smallest, based on values in table 6-3, e.g., 1 of 10 is equivalent to the largest system. "Radius of catchment" is the average extent of the system in space delimited by the 10% contour interval of the system (cf. table 6-3). The "Estimated number of consumers" is the estimated number of people within the total system (cf. table 6-3). The "Area within California" is the percentage of area within California inscribed by the radius of catchment. The "Symmetry" is the relative degree of similarity of form of the system on either side of a dividing plane. The options are symmetrical or asymmetrical. "Shape" is the abstract geometrical form of the system (e.g., circular, elliptical, square, or triangular). "Direction of greatest extension" is the direction in which the obsidian is exchanged over the greatest distance. "Direction of lateral compression" is the direction in which a given system is attenuated in space. "Index value at 40 km" is an arbitrary distance from the obsidian source to standardize the rate of decrease of obsidian utilizatin (cf. contour values on plates 6-1 through 6-10). "Maximum index value at source" is the value of the obsidian utilization index at the source itself. This value varies for some sources, since other lithic (non-obsidian) materials apparently are used in near proximity to the source. "Neighboring sources" are the next nearest sources (consult plate 4-2). "Presence of intersource competition" is a qualitative statement on the degree of competition between sources based upon the extent of lateral compression and deformation of the system relative to its neighbors. "Presence of alternative lithic source competition" is the degree to which major non-obsidian lithic sources appear to affect the system (cf. plate 5-4).

Data under "Ethnographic data" were derived from Davis (1961) and several other authors as noted in the following text.

Obsidian Butte Exchange System (Plate 6-1)

Source Data

Source included: Obsidian Butte (T11S, 13E, SBBM) perhaps unlocated Mexican source in Baja.

Location: Southern California.

Color of obsidian: black.

Crystallization: 10%.

Occurrence of material: blocks and rounded cobbles.

Hardness: 562 kg/mm^2.

Structure of the deposit: dome?

Area of the deposit: approximately 0.5 sq. km.

Elevation of the deposit: near sea level.

Surrounding parent rock: lacustrine sediment.

Age of the eruption: 18,000-55,000 years ago.

Present state of deposit: site of modern gravel and pumice quarry.

System Characteristics

Rank size: 9 of 10.

Radius of catchment: 159 km.

Estimated number of consumers: c. 21,000 persons.

Area within California: 70%.

Symmetry: asymmetrical.

Shape: "Triangular."

Direction of greatest extension: north.

Direction of lateral compression: none observed.

Index value at 40 km: 30%-50%.

Maximum index value at source: 60%-70%.

Neighboring sources: Coso to the north, Mexican source to the south.

Presence of intersource competition: little.

Presence of alternative lithic source competition: unknown.

Ethnographic Data

Group controlling source: Kamia.

Groups giving access to source: unknown.

Exchange items received by "source" group: (2) tobacco, acorns, baked mescal roots, yucca fiber sandals, baskets, eagle feathers, carrying nets, shells from Gulf of California.

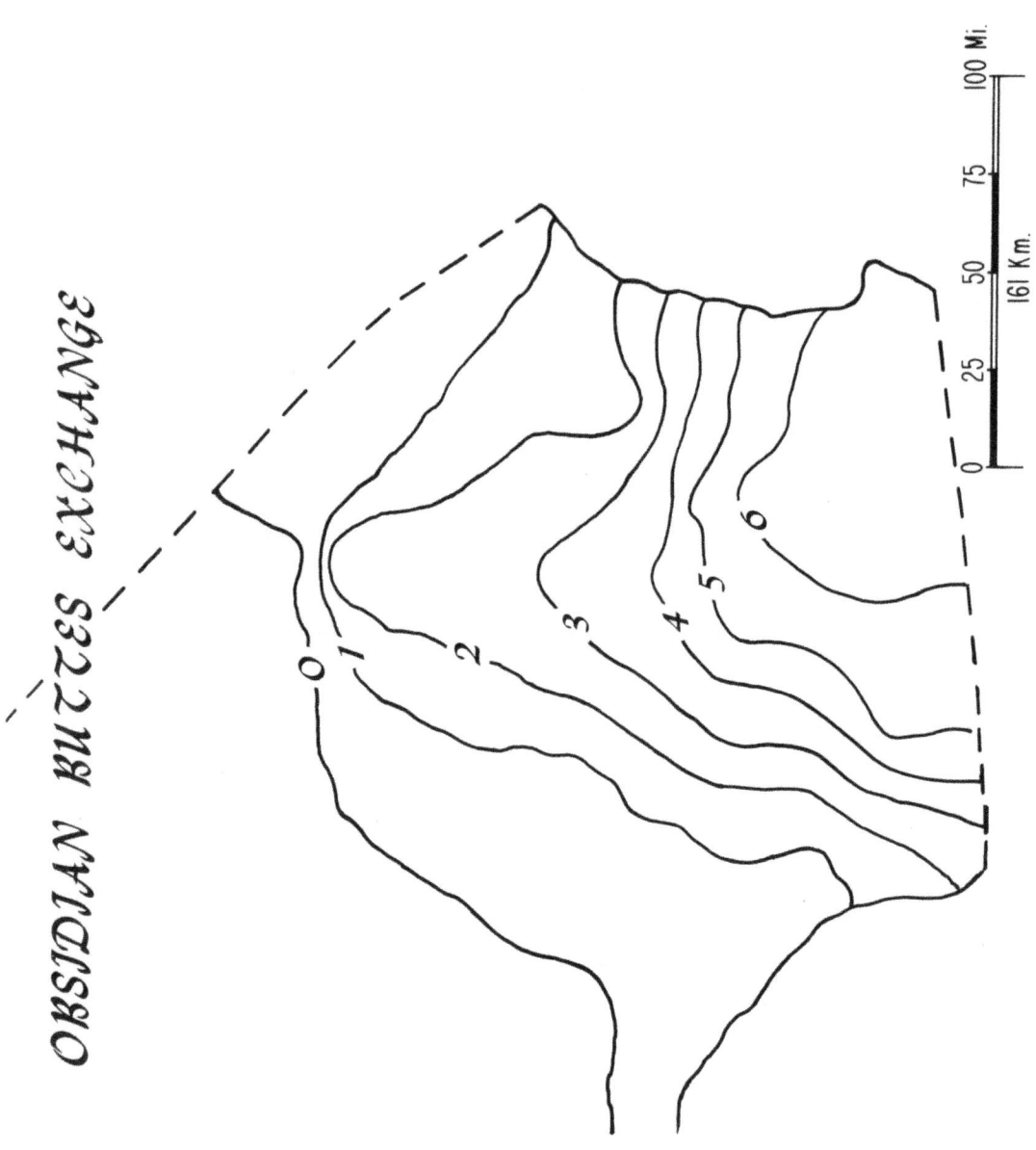

Plate 6-1. Obsidian Buttes Exchange. The archaeological distribution of this obsidian source is the result of the synagraphic mapping of data referred to in Table 5.1 and provided by Jack (1976). Each contour represents a 10% interval of the total chipped stone tool industry.

Obsidian export: not recorded.

Other items exported: no obsidian recorded, vegetal foods, salt, tobacco.

Primary receivers of obsidian: Chemehuevi, Cahuilla, Diegueño.

Percentage of All Obsidian Used (Jack 1976)

Not available.

Bodie Hills Exchange System (Plate 6-2)

Source Data

Source(s) included: Bodie Hills (T5N, R26E, MDBM), Pine Grove Hills (T9N, R26E, MDBM), Mt. Hicks (T5N, R29E, MDBM).

Location: East Central California.

Color of obsidian: clear gray and black.

Crystallization: 0%.

Occurrence of material: cobble-to-pebbles.

Hardness: 907 kg/mm^2.

Structure of the deposit: massive flows.

Area of the deposit: approximately 9 sq. km.

Elevation of the deposit: 2.1-2.6 km.

Surrounding parent rock: rhyolite.

Age of the eruption: unknown, Pleistocene?

Present state of deposit: little geologic material remaining.

System Characteristics

Size: major; 3 of 10.

Radius of catchment: 226 km.

Estimated number of consumers: c. 98,000 persons.

Area within California: 70%-80%.

Symmetry: asymmetrical.

Shape: elliptical and bimodal.

Direction of greatest extension: WSW.

Direction of lateral compression: NW and SE.

Index value at 40 km: 40%-50%.

Maximum index value at source: 90%-100%.

Neighboring sources: St. Helena and Annadel, west; Casa Diablo, south; Surprise Valley, north.

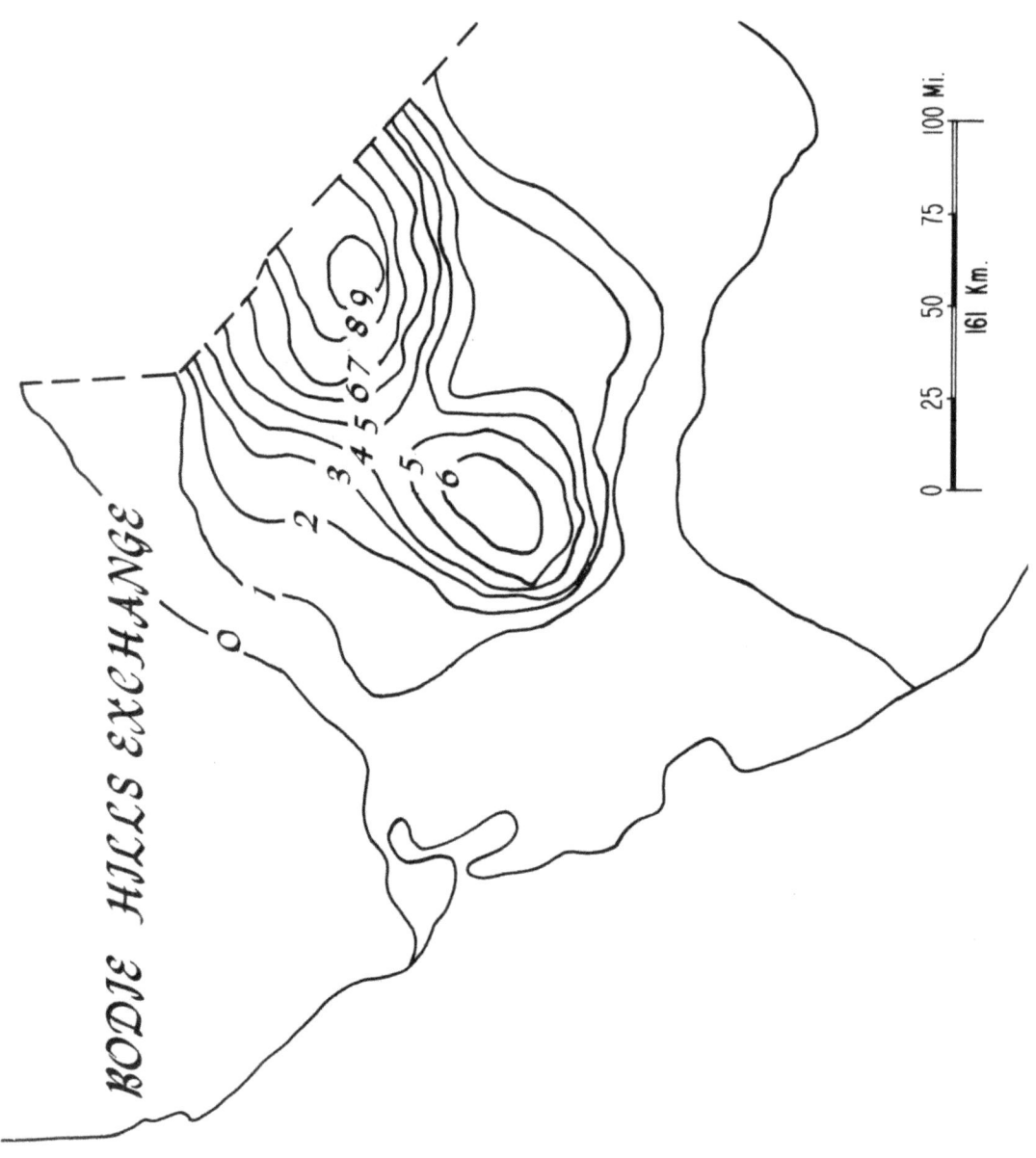

Plate 6-2. Bodie Hills Exchange. The archaeological distribution of this obsidian source is the result of the synagraphic mapping of data referred to in Table 5-1 and provided by Jack (1976). Each contour represents a 10% interval of the total chipped stone tool industry.

Presence of intersource competition: Casa Diablo and St. Helena.

Presence of alternative lithic source competition: unknown, probably not important.

Ethnographic Data

Group controlling source: Washo.

Groups giving access to source: none recorded.

Exchange items received by "source" group: (2) acorns, shell beads, seashells, baskets, papam bulbs, redbud bark for basketry, soaproot leaves for brushes, kutsavi, beads, shells, baskets, manzanita berries.

Obsidian export: not recorded.

Other items exported: (2) salt, piñon nuts, buffalo skin robes, rabbit skin blankets.

Primary receivers of obsidian: N. Paiute, Maidu, Eastern Mono, Western Mono.

Percentage of Total Obsidian Used (Jack 1976)

Washo: 8.5 (to the north of the source)
Bay Miwok: 3.4
Plains Miwok: 39.7
N. Sierra Miwok: 75
S. Sierra Miwok: 25.2
Western Mono: 2.7
Coast Miwok: 0.9
Costanoan: 5.5

Casa Diablo Exchange System (Plate 6-3)

Source Data

Source(s) included: Casa Diablo (T3S, R28E, MDBM), Queen (T1N, R32E, MDBM), Mono Glass Mt. (T1S, R30E, MDBM), Mono Craters (T1N, R27E, MDBM).

Location: East Central California.

Color of obsidian: black, gray, red streaks, very varied.

Crystallization: 14% (Casa Diablo).

Occurrence of material: cobble-pebbles (Casa Diablo).

Hardness: 799 kg/mm^2 (Casa Diablo).

Structure of the deposit: Coulee?

Area of the deposit: 15 sq. km (Casa Diablo).

Elevation of the deposit: 2.2-2.3 km.

Surrounding parent rock: volcanic.

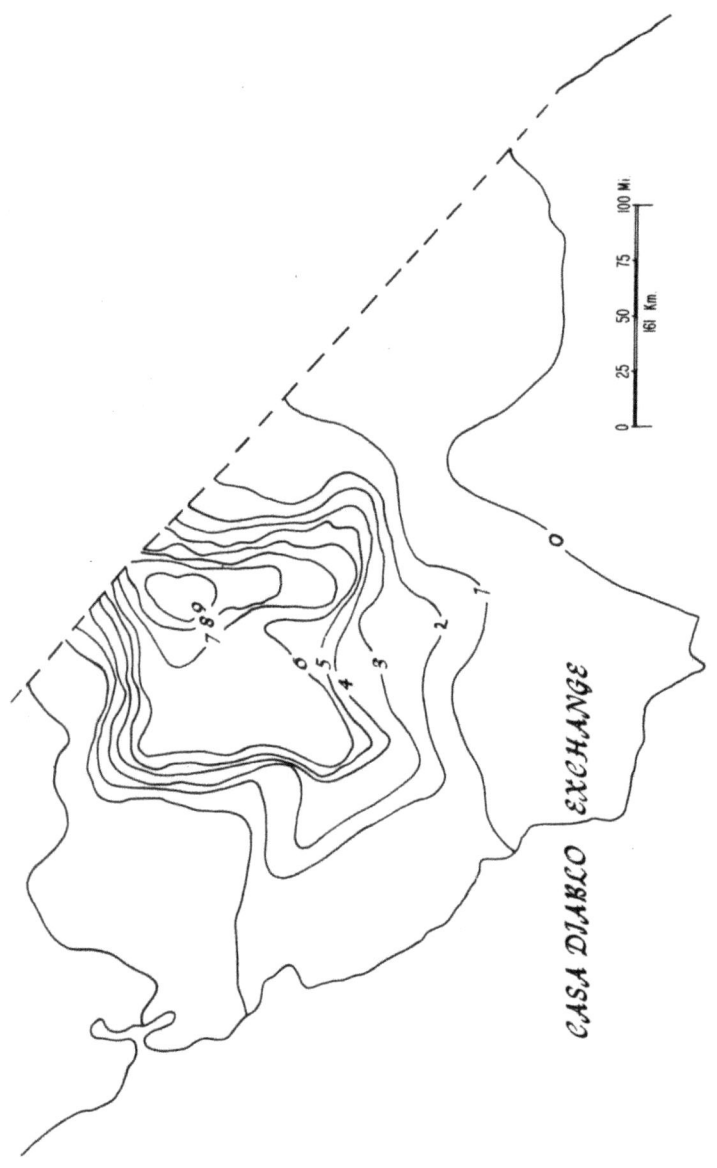

Plate 6-3. Casa Diablo Exchange. The archaeological distribution of this obsidian source is the result of the synagraphic mapping of data referred to in Table 5-1 and provided by Jack (1976). Each contour represents a 10% interval of the total chipped stone tool industry.

Age of the eruption: unknown (Casa Diablo), Mono Craters, 1,300-12,000 years ago, Mono Glass Mt. 0.9 million years ago.

Present state of deposit: geologic material remaining, many quarry workshops, protected by Forest Service.

System Characteristics

Size: major; 2 of 10.

Radius of catchment: 256 km.

Estimated number of consumers: c. 354,000 persons.

Area within California: 70%-90%.

Symmetry: asymmetrical.

Shape: "square" ?

Direction of greatest extension: south.

Direction of lateral compression: NW, SW, and E.

Index value at 40 km: 40%-60%

Maximum index value at source: 90%-100%.

Neighboring sources: Bodie Hills, north; Fish Springs, south; St. Helena, west.

Presence of inter-source competition: Bodie Hills.

Presence of alternative lithic source competition: chert from Coast Ranges.

Ethnographic Data

Group controlling source: Mono Lake Paiute (Eastern Mono).

Groups giving access to source: none recorded.

Exchange items received by "source" group: (2) squaw berries, (5) shell beads, (2) glass beads, (2) acorns, (4) baskets, manzanita berries, bear skins, (2) rabbit skin blankets, (3) elderberries, arrows, (2) clam disk beads, fungus used in paint, black paint, yellow paint, deer-antelope-elk skins, steatite, salt grass, salt, (2) acorn meal, fine Yokuts baskets, shell ornaments, buckskins.

Obsidian export: recorded as unfinished obsidian arrowheads.

Other items exported: (5) salt, (5) pine nuts, seed food, (3) rabbit skin blankets, tobacco, (2) baskets, buckskin, pottery vessels, clay pipes, pandora moth caterpillars, (2) kutsavi, (3) red paint, white paint, pumice stone, basketry materials, (2) sinew-bucked bowls, (2) moccasins, jerked deer meat, (2) hot-rock lifters, shell beads, mineral paint, pitch-lined basketry water bottles, acorns, mountain sheep skins, tailored sleeveless buckskin jackets, fox-skin leggings, unfinished obsidian arrowheads.

Primary receivers of obsidian: Washo, Western Mono, Koso, Owens Valley Paiute.

Percentage of Total Obsidian Used (Jack 1976)

    Panamint Shoshone: 20.6
    Owens Valley Paiute: (Iny-76) 20.9; (Iny-1, 2) 26.4
    Panamint Mountains: (Kawaiisu) 31.8
    Kings Canyon, Sierra: 75
    Plains Miwok: 9.5
    S. Sierra Miwok: 72.7
    Western Mono: 95.6
    N. Valley Yokuts: 97.6
    S. Valley Yokuts: 66.7
    Buena Vista Lake Yokuts: 18.5
    Costanoan: 5.5
    Note: Owens Valley Paiutes quarried at Mono Glass Mt. to the east.

Fish Springs Exchange System (Plate 6-4)

Source Data

    Source included: Fish Springs (T10S, R33E, MDBM).

    Location: East Central California.

    Color of obsidian: gray-black.

    Crystallization: 2%.

    Occurrence of material: pebbles (almost too small for projectile point manufacture).

    Hardness: 714 kg/mm^2.

    Structure of the deposit: pumice dome.

    Area of the deposit: 0.5 sq. km (approx.).

    Elevation of the deposit: 0.95 km.

    Surrounding parent rock: volcanics.

    Age of the eruption: unknown, Pleistocene?

    Present state of deposit: active pumice quarry, mining claim, potters from nearby campground.

    Note: source designated as "kapi or obsidian" by Owens Valley Paiute.

System Characteristics

    Size: minor; 10 of 10.

    Radius of catchment: 68 km.

    Estimated number of consumers: c. 17,000 persons.

    Area within California: 100%.

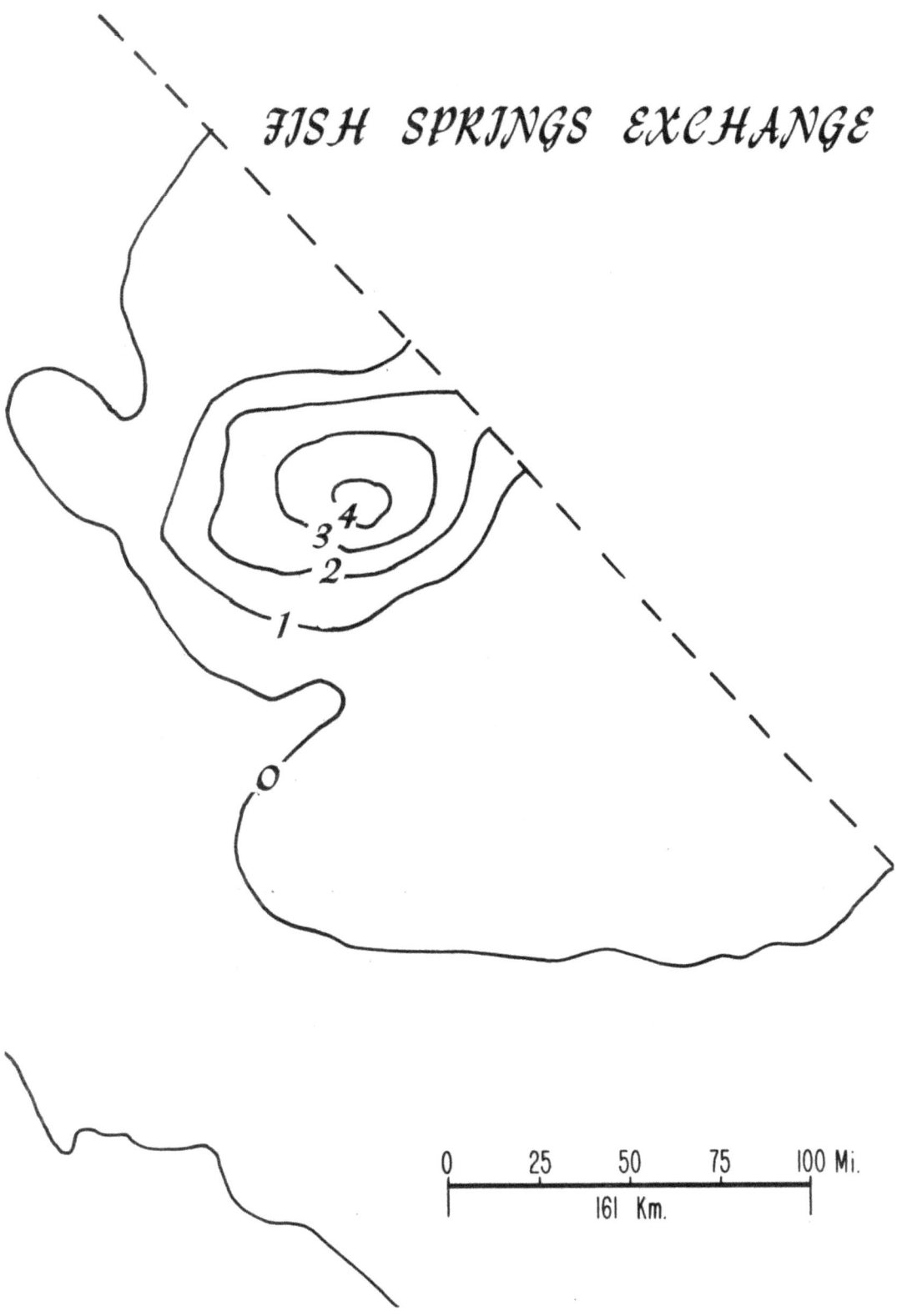

Plate 6-4. Fish Springs Exchange. The archaeological distribution of this obsidian source is the result of the synagraphic mapping of data referred to in Table 5-1 and provided by Jack (1976). Each contour represents a 10% interval of the total chipped stone tool industry.

Symmetry: symmetrical, yet eccentric.

Shape: circular.

Direction of greatest extension: slight NE-SW elongation.

Direction of lateral compression: slight NW and SW compression.

Index value at 40 km: 15%-25%.

Maximum index value at source: 40%-50%.

Neighboring sources: Casa Diablo, north; Coso, south.

Presence of inter-source competition: Casa Diablo and Coso.

Presence of alternative lithic source competition: unknown.

Ethnographic Data

Group controlling source: Owens Valley Paiute (Eastern Mono).

Groups giving access to source: unknown.

Exchange items received by "source" group: see Casa Diablo.

Obsidian exported: difficult to determine.

Other items exported: see Casa Diablo.

Primary receivers of obsidian: Mono Lake Paiute, Western Mono, Koso.

Percentage of Total Obsidian Used (Jack 1976)

Panamint Shoshone: 5.9
Owens Valley Paiute: (Iny-76) 70.08; (Iny 1, 2) 5.3
Panamint Mountains: (Kawaiisu) 31.8
Kings Canyon, Sierra: 25.0

Note: this source designated as "kapi or obsidian" by Owens Valley Paiute.

Annadel Exchange System (Plate 6-5)

Source Data

Source(s) included: Annadel (T17N, R6W, MDBM).

Location: West Central California.

Color of obsidian: black.

Crystallization: 88%.

Occurrence of material: pebbles (almost too small for projectile point manufacture).

Hardness: 590 kg/mm^2.

Structure of the deposit: unknown.

Area of the deposit: unknown.

Elevation of the deposit: 0.3 km.

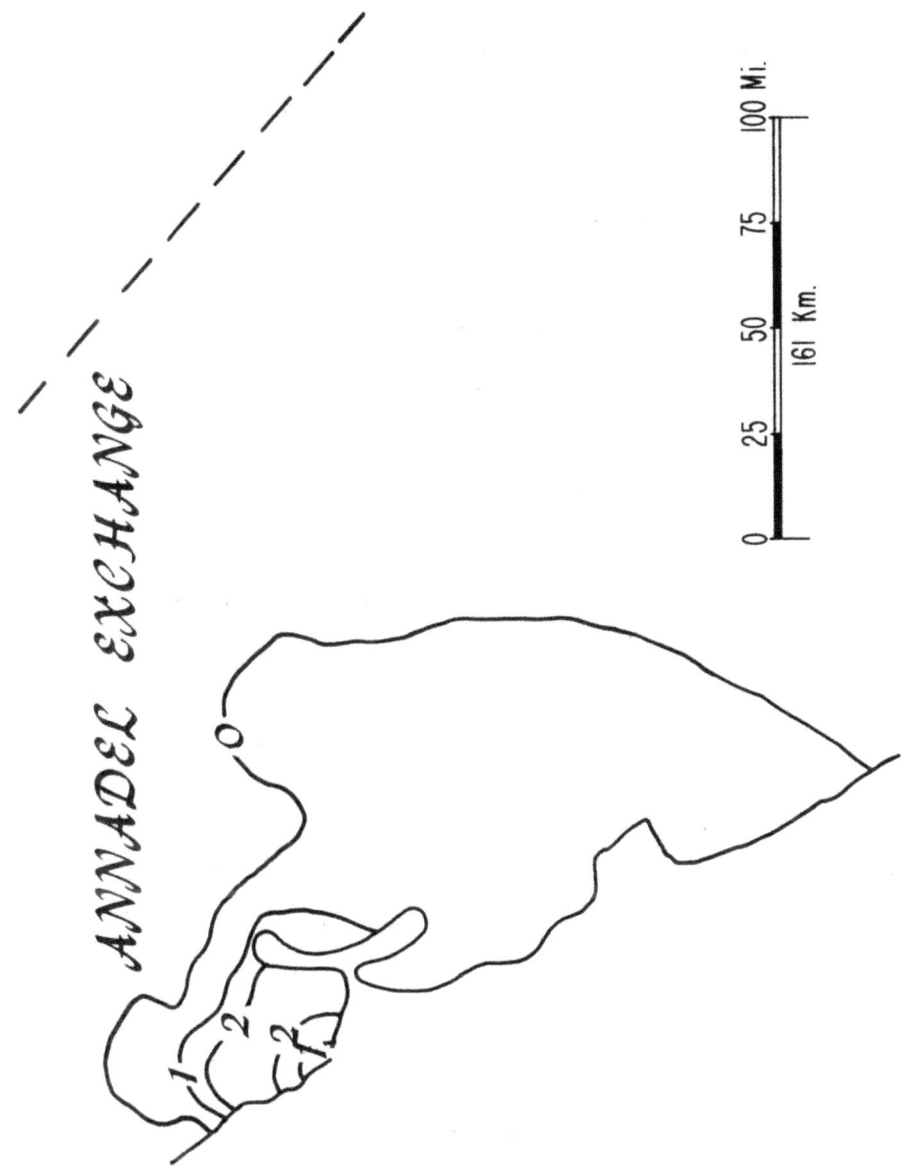

Plate 6-5. Annadel Exchange. The archaeological distribution of this obsidian source is the result of the synagraphic mapping of data referred to in Table 5.1 and provided by Jack (1976). Each contour represents a 10% interval of the total chipped stone tool industry.

Surrounding parent rock: volcanic.

Age of the eruption: unknown.

Present state of deposit: administered County Park.

System Characteristics

Size: minor; 8 of 10.

Radius of catchment: 98 km.

Estimated number of consumers: 30,000 persons.

Area within California: 100%.

Symmetry: asymmetrical.

Shape: elliptical.

Direction of greatest extension: E-W.

Direction of lateral compression: NE.

Index value at 40 km: 0%-10%.

Maximum index value at source: 20%-30%.

Neighboring sources: St. Helena, northeast.

Presence of inter-source competition: St. Helena.

Presence of alternative lithic source competition: chert in Coast Ranges.

Ethnographic Data

Group controlling source: Coast Miwok.

Groups giving access to source: none recorded.

Exchange items received by "source" group: no record.

Obsidian export: none recorded.

Other items exported: (2) clam shells, abalone shells, clam disk beads.

Primary receivers of obsidian: Wappo, Pomo, Costanoan, Bay Miwok.

Percentage of Total Obsidian Used

Coast Miwok: 39.8
Costanoan: 5.5
Plains Miwok: 1.6

St. Helena Exchange System (Plate 6-6)

Source Data

Source(s) included: St. Helena (or Napa Glass Mt.) (T8N, R6W, MDBM).

Location: West Central California.

Color of obsidian: black, opaque.

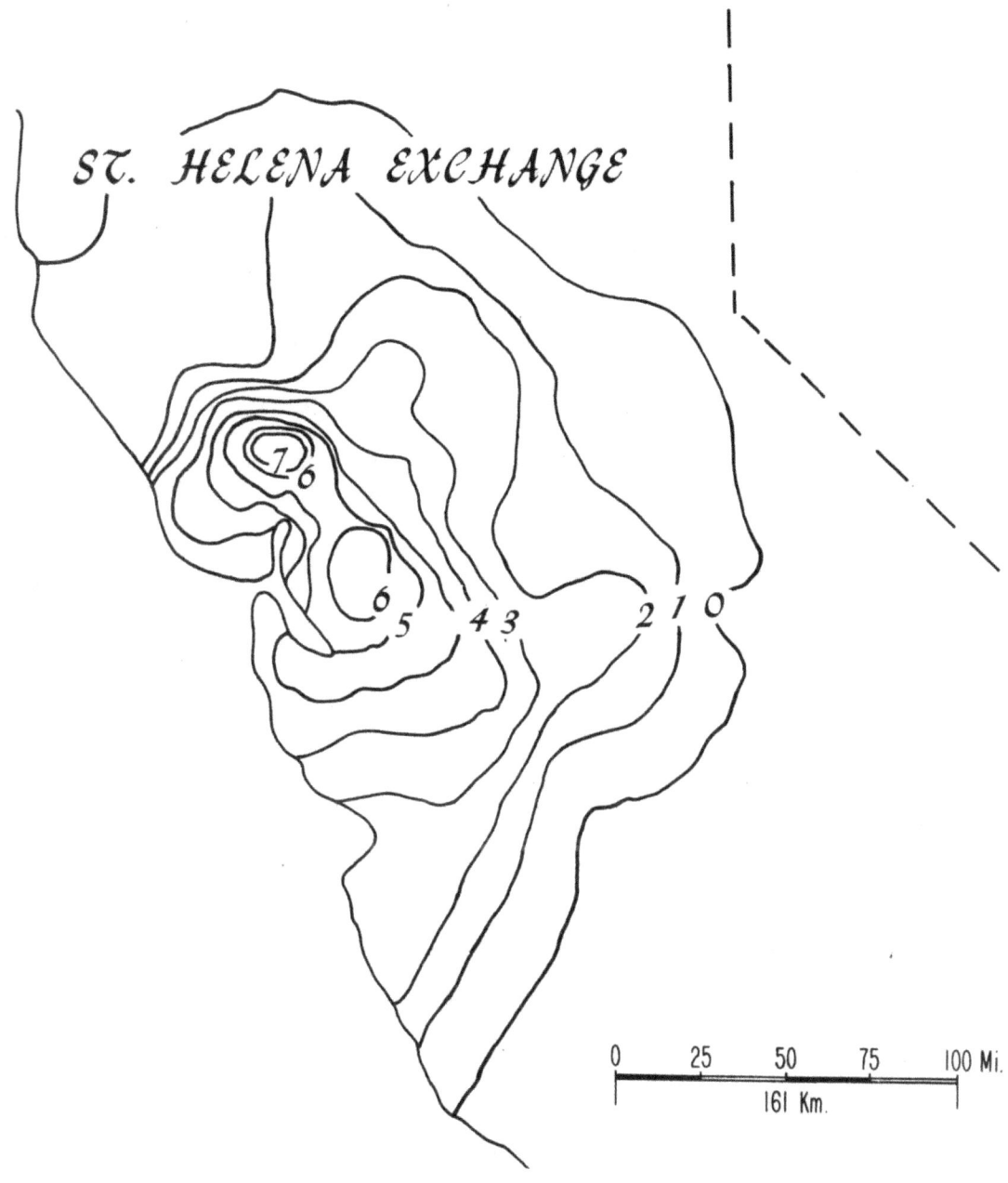

Plate 6-6. St. Helena Exchange. The archaeological distribution of this obsidian source is the result of the synagraphic mapping of data referred to in Table 5-1 and provided by Jack (1976). Each contour represents a 10% interval of the total chipped stone tool industry.

Crystallization: 9%.

Occurrence of material: blocks, bombs, cobbles, pebbles.

Hardness: 789 kg/mm^2.

Structure of the deposit: dome-like or faulted sill.

Volume of the deposit: 100,000 cubic ft. (Heizer and Treganza 1944).

Elevation of the deposit: 0.1 km.

Surrounding parent rock: volcanic.

Age of the eruption: unknown.

Present state of deposit: protected by poison oak, rural neighborhood, and land under development.

System Characteristics

Size: major; 1 of 10.

Radius of catchment: 206 km.

Estimated number of consumers: c. 392,000 persons.

Area within California: 100%.

Symmetry: asymmetrical.

Shape: elliptical and bimodal.

Direction of greatest extension: SE and NE.

Direction of lateral compression: NW.

Index value at 40 km: 30%-60%.

Maximum index value at source: 70%-80%.

Neighboring sources: Clear Lake, north; Annadel, southwest; Bodie Hills, east; Casa Diablo, southeast.

Presence of inter-source competition: all the above.

Presence of alternative lithic source competition: possibly chert in Coast Ranges.

Ethnographic Data

Group controlling source: Wappo.

Groups giving access to source: unknown.

Exchange items received by "source" group: (2) sinew-backed bows, clam disk beads, clam shells, abalone shells, tule mats, fish, magnesite beads, yellow hammer headbands, clams.

Obsidian export: none recorded.

Other items exported: salt.

Primary receivers of obsidian: Lake Miwok, Coast Miwok, Wintun, Pomo.

Percentage of Total Obsidian Used

    Pomo: 8.5
    Wappo: 100
    Hill Wintun: 15.6
    River Wintun: 63.0
    Coast Miwok: 55.9
    Costanoan: 81.5
    Bay Miwok: 96.6
    Plains Miwok: 47.6
    N. Sierra Miwok (+ Cal-82) 25.0

Clear Lake Exchange System (Plate 6-7)

Source Data

Source(s) included: Borax Lake (T13N, R7W, MDBM), Mt. Konocti (T13N, R8W, MDBM).

Location: West Central California.

Color of obsidian: black; black and gray banded.

Crystallization: 4% (Borax Lake); 10% (Mt. Konocti).

Occurrence of material: cobbles and blocks.

Hardness: 734 kg/mm^2 (Borax Lake; 541 kg/mm^2 Mt. Konocti).

Structure of the deposit: sill, flows.

Area and volume of the deposit: 2.6 sq. km, hundreds of thousands of cubic feet debitage (Anderson 1936).

Elevation of the deposit: 0.3 km.

Surrounding parent rock: varied.

Age of the eruption: Pleistocene greater than 10,000 years B.P. (Ericson and Berger 1974).

Present state of deposit: active construction zone.

System Characteristics

Size: major; 4 of 10.

Radius of catchment: 138 km.

Estimated number of consumers: c. 110,000 persons.

Area within California: 100%.

Symmetry: asymmetrical.

Shape: elliptical and bimodal.

Direction of greatest extension: SW.

Direction of lateral compression: SE and NW.

Index value at 40 km: 10%-60%.

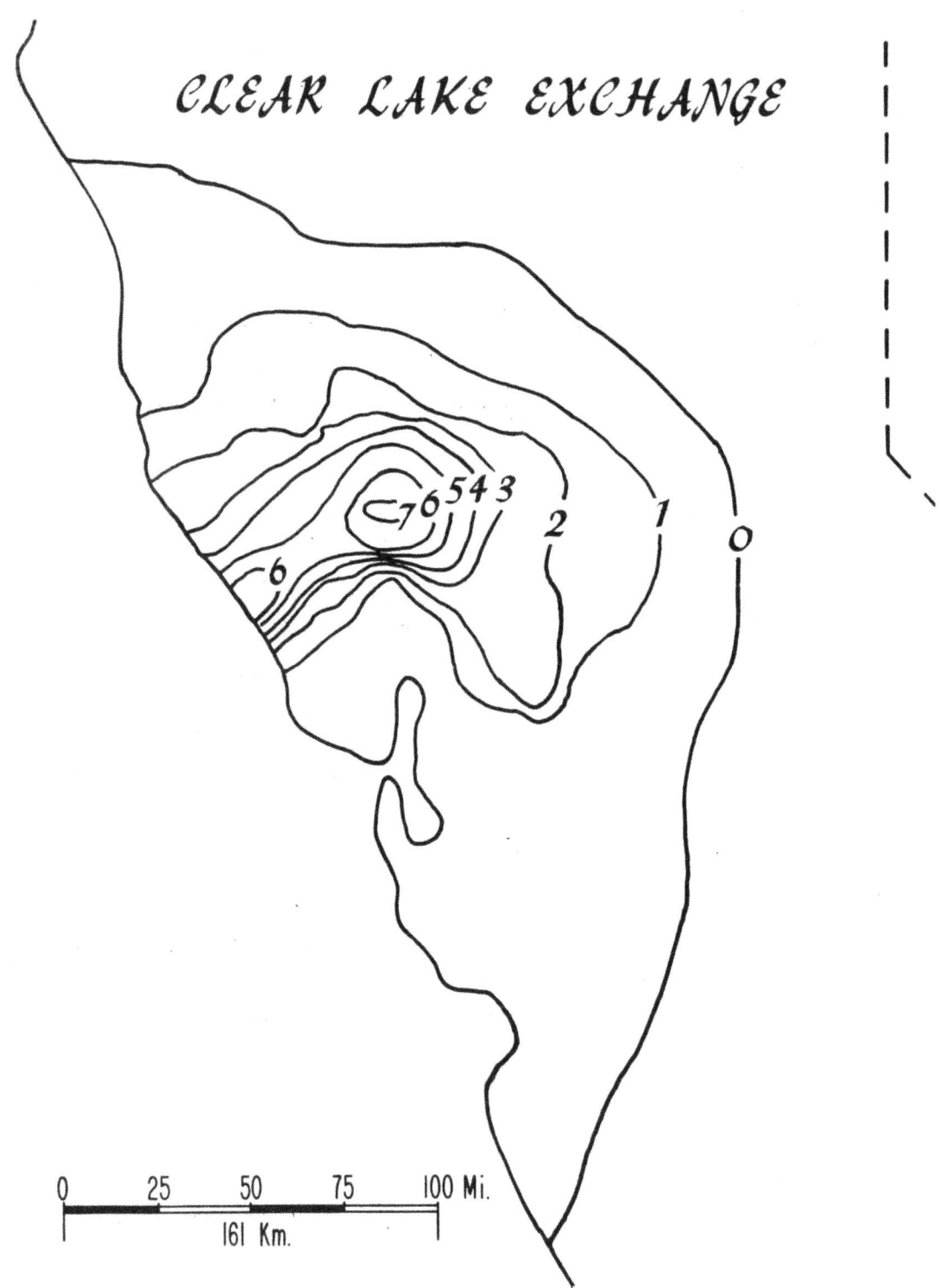

Plate 6-7. Clear Lake Exchange. The archaeological distribution of this obsidian source is the result of the synagraphic mapping of data referred to in Table 5-1 and provided by Jack (1976). Each contour represents a 10% interval of the total chipped stone tool industry.

Maximum index value at source: 70%-80%.

Neighboring sources: Medicine Lake, Northeast; St. Helena, South.

Presence of inter-source competition: St. Helena.

Presence of alternative lithic source competition: chert in Coast Ranges.

Ethnographic Data

Group controlling source: Pomo.

Groups giving access to source: Masut group of Pomo, Long Valley Wintun, Coyote Valley Miwok.

Exchange items received by "source" group: furs, beads, blankets, skins, Iris fiber cord for deer shanes, arrows, (2) sinew-backed bows of yew, yellow hammer headbands, woodpecker scalp belts, cordage for making deer nets, surf fish, abalone, giant chiton, seaweed, salt, (2) magnesite beads, shell beads, acorns, tule mats, sinew-backed bows, fish.

Primary receivers of obsidian: Huchon, Coast Miwok, Coast Yuki, Wappo, Wintun, Lake Miwok.

Percentage of Total Obsidian Used (Jack 1976)

Coast Yuki: x
Pomo: 91.5
Lake Miwok: x
Hill Wintun: 84.4
River Wintun: 37.0
Coast Miwok: 3.4
Plains Miwok: 1.6

Notes: Pomo of Clear Lake divided local obsidian into two types: "bati xaga", "arrow-obsidian" from Borax Lake, and "dupa xaga", "to cut obsidian" from Cole Creek (Mt. Konocti) used for razors and knives because it breaks with sharp edges (Loeb 1936:152).

Medicine Lake Exchange System (Plate 6-8)

Source Data

Source(s) included: Medicine Lake (or Modoc Glass Mt.) (T44N, R5E, MDBM).

Location: Northern California.

Color of obsidian: gray.

Crystallization: 45%.

Occurrence of material: block, boulders, cobbles.

Hardness: 742 kg/mm^2.

Structure of the deposit: flow structure.

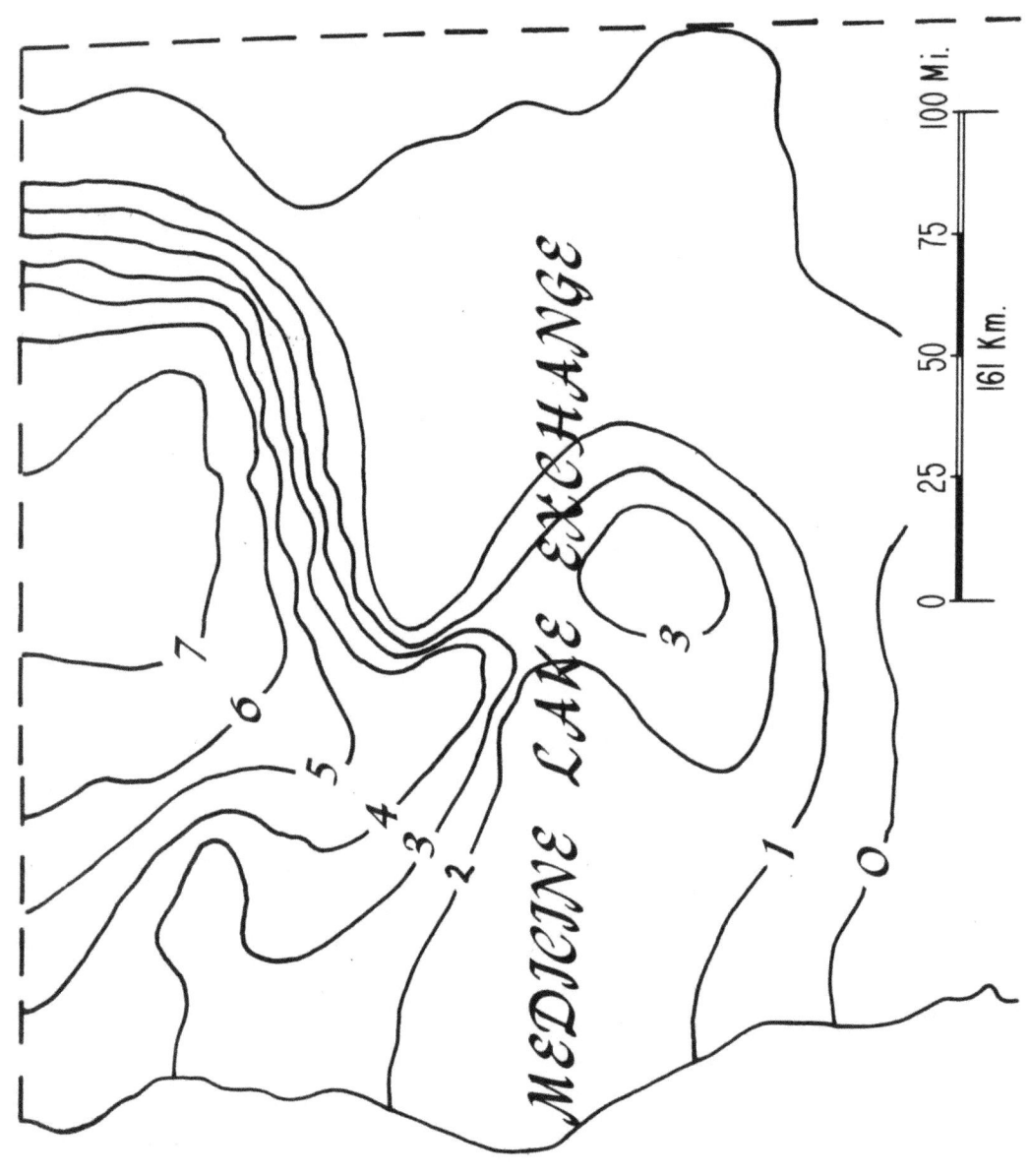

Plate 6-8. Medicine Lake Exchange. The archaeological distribution of this obsidian source is the result of the synagraphic mapping of data referred to in Table 5-1 and provided by Jack (1976). Each contour represents a 10% interval of the total chipped stone tool industry.

Area of the deposit: 93 sq. km (approx.).

Elevation of the deposit: 2.2-2.7 km.

Surrounding parent rock: volcanic dacite and rhyolite.

Age of the eruption: 190 ± 200 years for rhyolite flow (W-1546).

Present state of deposit: open to potting, access by road.

System Characteristics

Size: minor; 7 of 10.

Radius of catchment: 209 km.

Estimated number of consumers: c. 31,000 persons.

Area within California: 70%-80%.

Symmetry: asymmetrical.

Shape: "S-shaped."

Direction of greatest extension: SSW.

Direction of lateral compression: E.

Index value at 40 km: 30%-60%.

Maximum index value at source: 70%-80%.

Neighboring sources: Surprise Valley, east; Clear Lake, southwest; many obsidian sources to north and east in Oregon.

Presence of inter-source competition: Surprise Valley.

Presence of alternative lithic source competition: chert to the west in Coast Ranges.

Ethnographic Data

Group controlling source: Modoc.

Groups giving access to source: Atsugewi, Achomawi, Yana, McCloud River Wintun (Heizer and Treganza 1944; Kniffen 1926:297; Voeglin 1942:47).

Exchange items received by "source" group: shell beads, twined baskets, grass skirts, pine nuts, string skirts, bows, dentalia, women slaves, various hides, wooden war clubs with stone or bone inserts, grooved stone axes, feather blankets.

Obsidian export: none recorded.

Other items exported: human slaves, twined baskets, blankets, beads, clothing, axes, spears, fishhooks, furs, bows, dentalia, horses, buckskin jackets and coats.

Primary receivers of obsidian: Shasta, Chimariko, New River, Konomita, Okwanuchu, Achomawi, Atsugewi, Northern Paiute.

Percentage of Total Obsidian Used (Jack 1976)

    Maidu: 5.6
    Central and North Wintun: 74.0
    Northwest Coast: 85.7

Notes: Sacramento Valley Wintun split off blocks by building fire against it (Dubois 1935); same group did not quarry obsidian exposed to sun (Fowke 1896:57). Western Achomawi inspected flakes whether usable or not (Voeglin 1942:47). The Wintun myth: "Theft of obsidian," recorded by Dubois and Demetracoupoulou (1931:279), describes a volcanic eruption (Ericson, Hagan, and Chesterman 1976:229).

Surprise Valley Exchange System (Plate 6-9)

Source Data

Source(s) included: Buck Mt. (T44N, R15EMDBM), Surprise Valley (T45N, R14E, MDBM), Cowhead Lake (T47N, R17E, MDBM).

Location: Northeast California.

Color of obsidian: clear, black; brown and black mottled.

Crystallization: 59% (Sugarhill); 24% (Buck Mt.).

Occurrence of material: blocks, boulders, cobbles, pebbles.

Hardness: 601 kg/mm^2 (Sugarhill); 705 kg/mm^2 (Buck Mt.).

Structure of the deposit: flow structures, Buck and Sugarhill.

Area of the deposit: Approximately 50 sq. km, Buck and Sugarhill.

Elevation of the deposit: 2.1-2.4 km.

Surrounding parent rock: volcanic.

Age of the eruption: unknown, Pleistocene?

Present state of deposit: Buck Mt. brown exported to Japan and sold to rock shops for jewelry.

System Characteristics

Size: minor; 6 of 10.

Radius of catchment: 241 km.

Estimated number of consumers: c. 36,000 persons.

Area within California: 50%-70%.

Symmetry: asymmetrical.

Shape: elliptical and bimodal.

Direction of greatest extension: SW.

Direction of lateral compression: W and S.

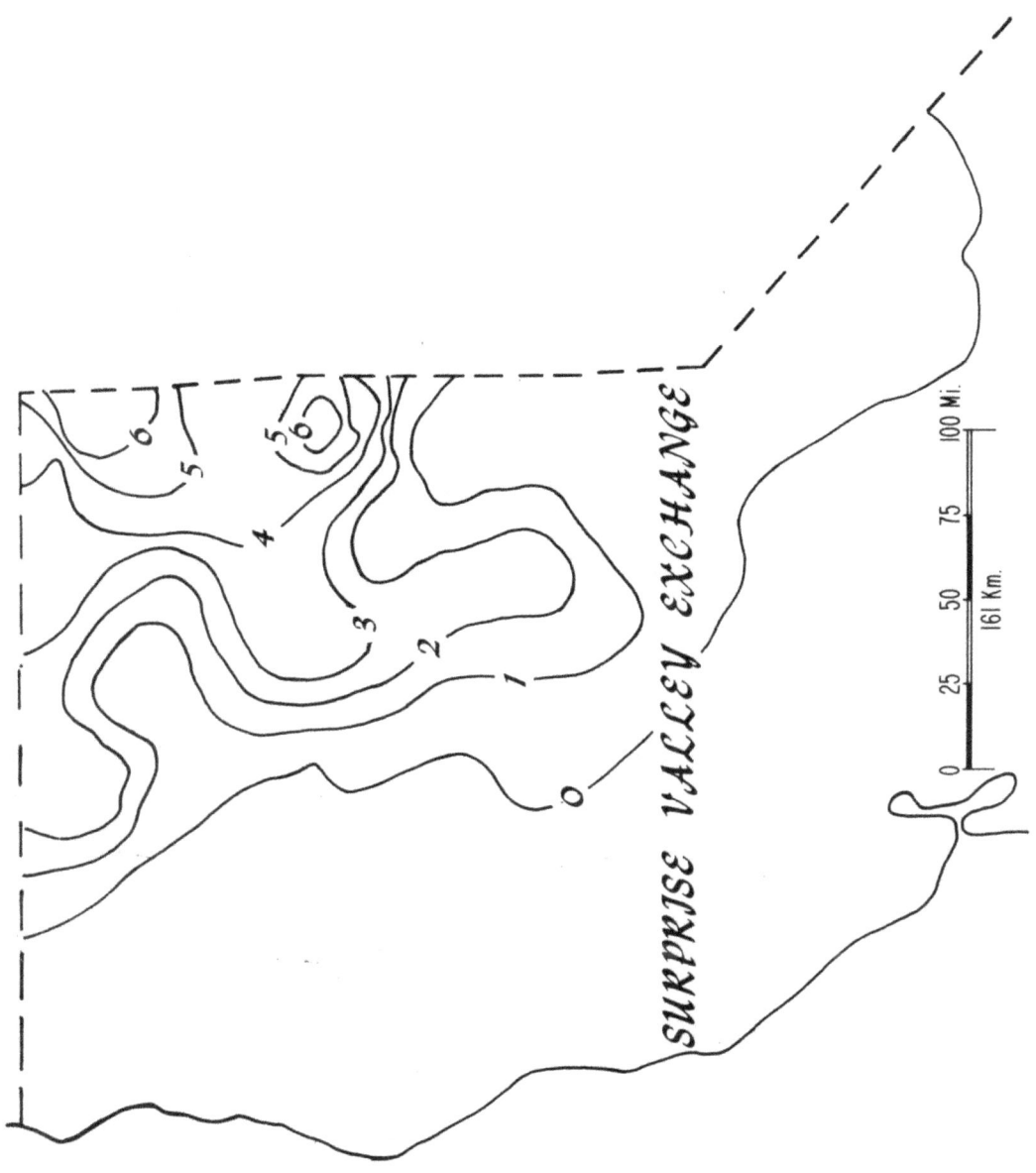

Plate 6-9. Surprise Valley Exchange. The archaeological distribution of this obsidian source is the result of the synagraphic mapping of data referred to in Table 5-1 and provided by Jack (1976). Each contour represents a 10% interval of the total chipped stone tool industry.

Index value at 40 km: 30%-60%.

Maximum index value at source: 60%-70%.

Neighboring sources: Medicine Lake, west; many Oregon sources to north and east.

Presence of inter-source competition: Medicine Lake.

Presence of alternative lithic source competition: unknown.

Ethnographic Data:

Group controlling source: Northern Paiute.

Groups giving access to source: Achomawi (Kniffen 1926:297).

Exchange items received by "source" group: sinew-backed bows, arrows, baskets, dried fish, women's basketry caps, clam disk beads, dried salmon flour, bows, basketry, shell beads, papam bulbs.

Obsidian export: none recorded.

Other items exported: basketry water bottles, sinew, arrowheads, red paint, buckskins, moccasins, rabbit skin blankets, various foods, hoes, buckskins, red oche, glass beads, guns, Olivella beads.

Primary receivers of obsidian: Achomawi, Atsugewi, Modoc, Maidu, Washo.

Percentage of Total Obsidian Used (Jack 1976)

Washo: 8.3
Maidu: 55.6
N. Paiute: 48.6
Central and Northern Wintun: 0.3

Note: Use of this source was a point of conflict between N. Paiute and Achomawi, since both groups claimed it (deAngulo and Freeland 1929: 313).

Coso Exchange System (Plate 6-10)

Source Data

Source(s) included: Coso (T22S, R38E, MDBM).

Location: Southeast California.

Color of obsidian: black.

Crystallization: 10%.

Occurrence of material: blocks and cobbles.

Hardness: 773 kg/mm^2.

Structure of the deposit: dome.

Area of the deposit: 2 sq. km.

Elevation of the deposit: 0.6 km.

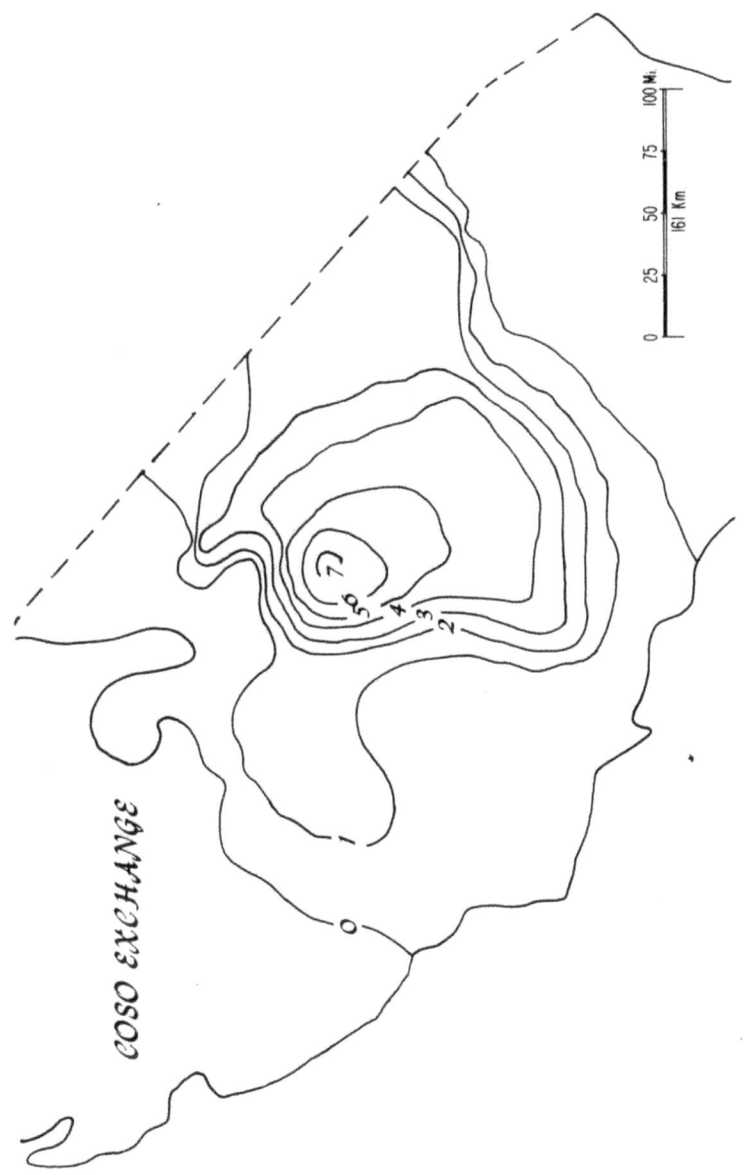

Plate 6-10 Coso Exchange. The archaeological distribution of this obsidian source is the result of the synagraphic mapping of data referred to in table 5-1 and provided by Jack (1976). Each contour represents a 10% interval of the total chipped-stone-tool industry.

Surrounding parent rock: volcanic.

Age of eruption: unknown, Pleistocene?

Present state of deposit: protected by U.S. Navy on China Lake Test Range.

System Characteristics

Size: major; 5 of 10.

Radius of catchment: 162 km.

Estimated number of consumers: c. 74,000 persons.

Area withn California: 95%.

Symmetry: asymmetrical.

Shape: "rectangular."

Direction of greatest extension: S.

Direction of lateral compression: W and SE.

Index value at 40 km: 15%-45%.

Maximum index value at source: 70%-80%.

Neighboring sources: Casa Diablo and Fish Springs to north, Obsidian Butte to south.

Presence of inter-source competition: little.

Presence of alternative lithic source competition: unknown.

Ethnographic Data

Group controlling source: Coso.

Groups giving access to source: Owens Valley Paiute (Farmer 1937:7).

Exchange items received by "source" group: shell beads, various goods.

Obsidian export: none recorded.

Other items exported: salt.

Primary receivers of obsidian: Eastern Mono, Western Mono, Tubatulabal, Chemehuevi, Kawaiisu.

Percentage of Total Obsidian Used (Jack 1976)

Tubatulabal: 100.0
Kawaiisu: 100.0
Panamint Shoshone: 73.5
Owens Valley Paiute: (Iny-76) 8.3; (Iny 1, 2) 63.2.
Panamint Mountain: (Kawaiisu) 54.5
S. Valley Yokuts: 33.3
Buena Vista Lake Yokuts: 77.8

Notes: Steward (1933) stated that the Tübatulabal once held the territory in which the quarry is located. However, in recent times the Koso, Panamint, and Koso-Panamint came into territory from the north (Farmer 1937).

COMPARATIVE SYSTEMS ANALYSIS

In the last section, the characteristics of each regional exchange system and certain details regarding the obsidian source have been presented. The changes in these characteristics, noted among all sources, are important clues as to the fundamental structure and dynamic operation of these systems. By conducting a comparative systems analysis, the importance of these structural units becomes evident. At this stage of analysis, only the most general characteristics are discussed.

Variations in the morphological properties of the systems are perhaps the most important. The morphological properties are those defined as "size" and "shape."

Among the 10 exchange systems, there is considerable variation in these properties. Ninety percent of the systems vary in form and are not circular in shape. Ninety percent of the systems are asymmetrical due to internal variations within each system. Likewise, the sizes of the 10 systems vary in terms of their catchment and individual population. The variation in the morphological properties suggests that the internal dynamics and controlling variables differ for each system.

It is important to suggest several mechanisms which may result in the formation of the types of systems under analysis. For example, in the majority of noncircular systems, there is a particular direction of extension and lateral compression. The direction of extension is the direction along which the greatest quantity of the item is exchanged. Perpendicular to this direction may be the direction of attenuation, or lateral compression. Lateral compression may be due to a number of interesting conditions. For example, lateral compression may be the result of "competition" between two or more similar sources, the presence of parallel biomes (compression would be exhibited parallel to their boundaries), or the presence of a linear exchange system such as that proposed by Wright and Zeder (1977) where the system is dominated by the exchange of two utilitarian sources. The Bodie Hills and Casa Diablo systems exemplify lateral compression due to inter-source competition. This is further illustrated in the lack of source overlap in plate 6-11. The Bodie Hills system shown in plate 6-2 north of Mono Lake is compressed on its southeastern boundary. Likewise, the Casa Diablo system (plate 6-3), located south of Mono Lake, is compressed on its northwestern boundary. As these two systems issue out of the Mono Lake area, they are separate systems. On the other hand, the Medicine Lake and Surprise Valley systems, the northern systems in California, exemplify only lateral compression as shown in plates 6-8 and 6-9.

Although it appears that systems interact by having equal attenuation relative to each other, some systems are subsumed by more dominant systems. For example, the Fish Springs system, shown in plate 6-4, is

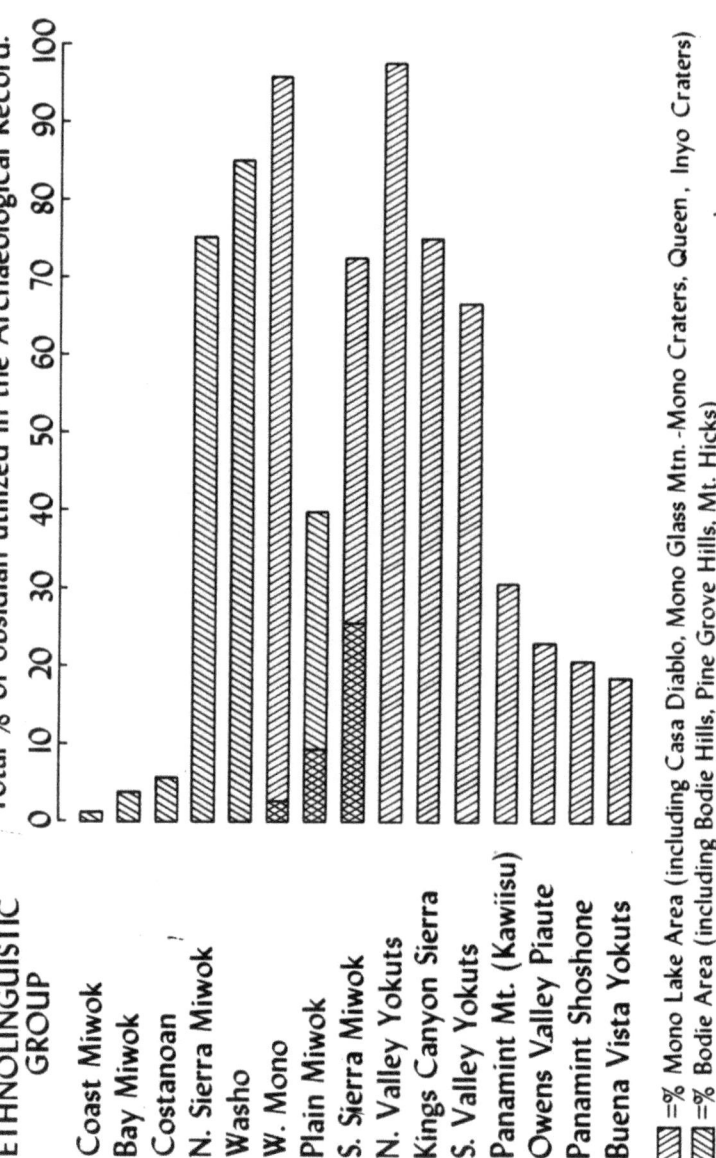

Plate 6-11. Spatial Separation of the Bodie Hills and Casa Diablo Exchange Systems. This histogram indicated the relative separation of the two systems based upon data from Jack (1976).

subordinate to the southern extent of the larger, dominant Casa Diablo system. This dominant-subordinate relationship is demonstrated by the inability of the Fish Springs system to laterally compress the Casa Diablo system, due to the low quantity of material utilized from this source and its restricted catchment.

Other important mechanisms may control the size of systems which should bear further investigation. One might expect that through similarities in the modes of production and transportation, systems would tend to have the same catchment; however, there appears to be a considerable variability in the catchment and estimated number of consumers within each system. This fact is more evident when comparing the catchment size of the systems analyzed later. It is extremely important to note that the variability in catchment does suggest that distance may not be a systemic control variable as has been suggested by Renfrew and others in earlier work (Renfrew, Dixon, and Cann 1968). These findings suggest the importance of evaluating several systemic control variables.

DISTANCE AS A VARIABLE

"Distance" from a source to point of consumption has been used as a variable to describe the change in consumption of an exchanged item (Renfrew, Dixon, and Cann 1968; Renfrew 1969; Hodder 1974). However, the significance of this variable has not been examined for two cases: (1) as a variable for the local distribution of an item, and (2) as a variable for trade at great distances. Both cases will be discussed here.

Locational geographers in formulating a spatial demand cone around a retail store (Berry 1967:61) assume an isotrophic plane (Garner 1968:307). It is also assumed that rational consumers respond to a store as a probability function of initial cost of an item plus their travel cost. Since in our society travel cost increases as a function of distance (and time), which can be measured in various ways (Haggett 1966:37), it is predicted that the distribution of an item obtained by consumers will decrease with distance (Berry 1967:61). For contemporary USA, distance appears to be a good estimator of travel cost determining the distribution of an item in space, holding fixed the means of transportation.

Distance remains an established measure of diminishing supply. For the analyst, it is a simple measure of the basic trends observed for most exchange systems. It is not clear, however, as to what "distance" is really measuring within a system or its cultural significance. The evaluation of how much variability of supply is explained by distance is vital to our further understanding of exchange systems. To start, a basic differentiation was made in chapter 4 between modes of acquisition. If an item is obtained by direct access, then it is assumed that distance can be equated with work expenditure in acquiring an item. Alternatively, if an item is obtained through exchange, the conversion of distance as a measure of work expenditure may be inappropriate. For example, in a simulated binary system of salt and ax exchange, the middlemen receive both items without much effort (Wright and Zeder 1977). The real work performed by the middlemen is minimal—only that expended in traveling between trade partners. As previously developed in chapter 4, in effect, the energy expenditure within the system is

Table 6-1

POPULATION ESTIMATES OF ETHNOLINGUISTIC GROUPS

Group	Population Estimate*	Group	Population Estimate*
Tolowa[3]	3.06	Poso Creek Yokuts[4]	2.48
Yurok[4]	4.18	S. Valley Yokuts[3]	7.12
Karok[4]	2.56	N. Valley Yokuts[3]	3.67
Wiyot[4]	10.76	Buena Vista Yokuts[4]	0.92
Hupa[3]	4.66	W. Mono[3]	5.05/5.85
Whilkut[4]	5.58	Costanoans[2]	1.83
Mattole[4]	5.47	Salinans[2]	0.88
Lolangkok Sinkyone[4]	8.16	Chumash[5]	2.02
Wailaki[3]	7.28	Wintun[2]	1.24
Coast Yuki[3]	4.52	NE Maidu[1]	0.78
Yuki[4]	5.88	Maidu (remaining)[2]	1.00
N. Pomo[4]	5.87	Yana[2]	0.94
E. Pomo[4]	4.95	Achomawi[1]	0.46
C. Pomo[4]	4.96	Atsugewi[1]	0.46
SE Pomo[4]	5.18	Chilula[3]	3.90
SW Pomo[4]	5.39	Chimariko[1]	0.61
Wappo[4]	8.86	Shasta[2]	1.27
Lake Miwok[4]	9.66	Lassik[3]	2.71
Plains Miwok[4]	11.18	Huchnom[3]	6.71
Coast Miwok[3]	3.08	Kato[1]	1.95
N. Sierra Miwok[4]	3.03	Modoc[1]	0.20
C. Sierra Miwok[4]	3.03	Esselen[1]	1.30
S. Sierra Miwok[3]	5.05	Washo[1]	0.23
N. Hill Yokuts[4]	10.75	N. Paiute[1]	0.17
Kings River Yokuts[4]	7.76	E. Mono[1]	0.32
Tule-Kaweah Yokuts[4]	5.61	Tubatulabal[3]	0.63
Coso[1]	0.06	Chemehuevi[1]	0.06
Kawaiisu[3]	0.63	Serrano[1]	0.29
Vanyume[1]	0.29	Kitanemuk[3]	0.63
Alliklik[3]	0.63	Gabrielino[1]	1.65
Fernandeño[1]	1.65	Luiseño[1]	1.54
Juaneño[1]	2.24	Cupeño[1]	1.35
Cahuilla[1]	0.81	Diegueño[1]	0.50
Kamia[1]	0.42	Mohave[1]	4.89
Halchidhoma	0.83	Yuma[1]	2.14

* People per square mile.
1. Kroeber 1925; 2,3. Cook 1943, 1955a, 1955b; 4. Baumhoff 1963; 5. Brown 1967.

much less than that expended by the same population using direct access, making the exchange system more efficient. Observed changes within the system must be related to the operation of other variables.

The dependency of a given group on an exchange item is an important consideration. If some fixed amount of an item is absolutely necessary for a given group or area, i.e., "critical" and without substitute, then it is expected that the quantity of the item remains constant relative to distance and can be directly related to the number of consumers. In this case, the quantity of the item is independent of distance. This relationship (constant value) seems to be demonstrated by the existence of the "supply zones" observed in many exchange systems. Renfrew, Dixon and Cann (1968) define a "supply zone" (bounded by greater than 80% obsidian) and a "contact zone" (less than 80% obsidian in the chipped stone tool category) for their Anatolian obsidian study. Similar "supply zones" have been shown for the distribution of Neolithic Cornish pottery and Neolithic Group VI axes (Hodder 1974:fig. 19). Ericson (1973b) considered that the extent of the "supply zone" was independent of distance and dependent on the availability of alternative materials. The boundary between the "supply and contact zones" (Renfrew, Dixon, and Cann 1968) most likely is determined by the availability of substitute or alternative materials of the same class or change in use of the obsidian. There is a definite change in the functional use of obsidian within the "supply and contact zones" as shown in plate 6-12.

Within a "contact zone" an item is considered here as "utilitarian." The amount of an item at a point is most likely related in some way to the number of consumers, i.e., that the amount of an item or supply decreases as some function of the number of consumers. A model of this relationship is as follows:

$$Q = aP^b \qquad [6-1]$$

Where Q is the amount of material, P is the integrated population, and a and b are constants. It is expected that the integrated population or the effective population of consumers between the source and any given point may be an unobserved variable which was only indirectly measured by distance in the Renfrew, Dixon, and Cann study (1968). The potential number of consumers with a system may be an important variable of an exchange system. Changes of population in space or time may have effects on the supply. For example, zones of high population density would tend to deplete the number of utilitarian items. In such areas, the item would be greatly attenuated with distance, assuming a fixed or inflexible production at the source. Most likely, a system can accommodate this increased demand by an increased production of obsidian at its source or restricting its use to certain tool categories in plate 6-12. The relationship between the number of consumers and diminishing supply warrants further examination. In summary, it is suggested that significance of distance as a variable depends on the mode of acquisition and the local importance of a resource. In direct access, distance can be related to work expenditure; however, distance within regional exchange systems has a different connotation: quite possibly, distance measures the effect of the cumulative number of consumers.

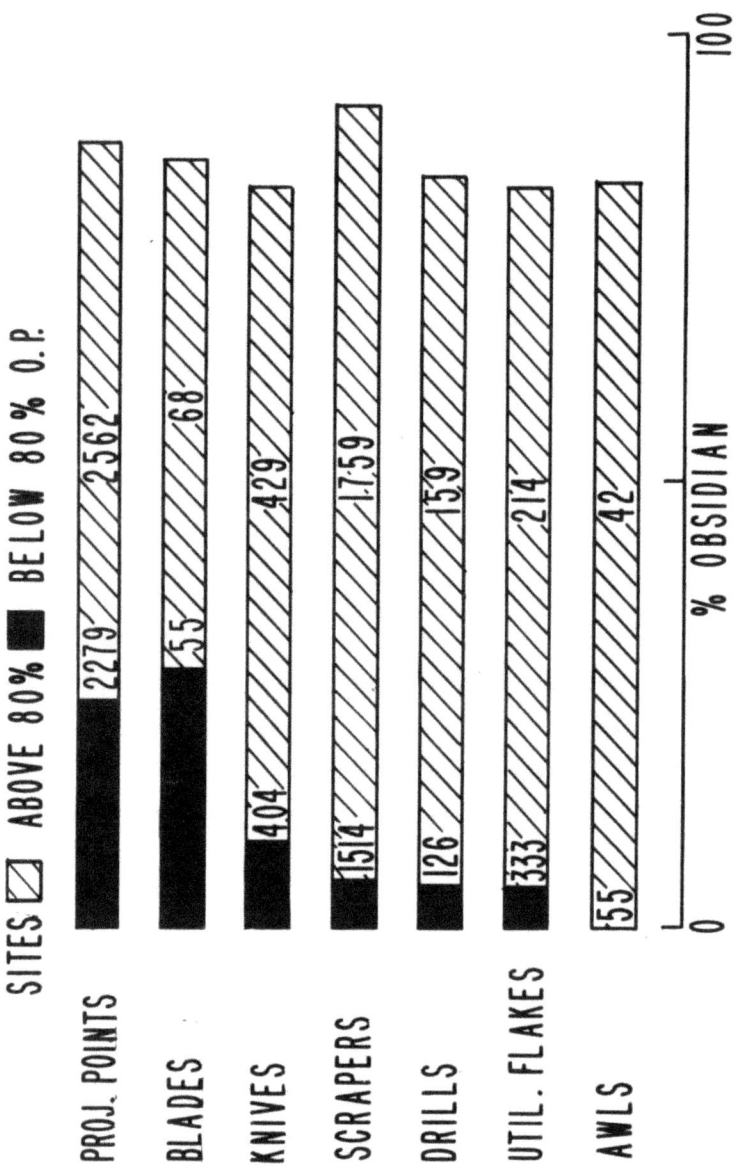

Plate 6-12. Differential Utilization of Lithic Materials for Making Stone Tools within the "Supply Zones" and the "Contact Zones," as defined by Renfrew et al. 1968. This data was compiled from the site records of 52 California archaeological sites, which specified the type of lithic materials used for particular tools. This histogram indicates that there is selective utilization of obsidian for producing specific tool types, mainly projectile points and knives, within the "Contact Zone."

QUANTITATIVE SYSTEMS ANALYSIS

In the following analysis, an empirical model of interrelationships of variables is evaluated. In this model, the simultaneous effects (1) effective population, (2) distance, and (3) distance to the nearest alternative obsidian source on the exchange index are evaluated for the 10 individual exchange systems. The distances from the point of observation to the source and next nearest source were measured. The ethnographic estimates of population density (Kroeber 1925; Cook 1943, 1955a, 1955b; Baumhoff 1963; Brown 1967) were used to calculate the effective population or number of consumers tabulated in table 6-1. The population within each ethnolinguistic segment was calculated by multiplying the respective population density, distance segment, and a 10-mile width, shown in plate 6-13. The values of each segment were added to determine the effective population. The above procedure generated the data forming the source-specific matrices of the exchange index, source distance, effective population, and alternative source distance.

The next step in the procedure was to define the mathematical model which was to be analyzed. The model, described in equation 6-2, assumes that the quantity of exchange item at any point is a function of the three variables where Q is the exchange index, <u>or percent of</u> a specific source of obsidian in the chipped-stone-tool category, P is the effective population, X is the distance to the source, and Y is the distance to the next nearest obsidian source.

$$Q = Q (P, X, Y) \qquad [6-2]$$

The linear model was selected for analysis to evaluate the simultaneous effects of the three variables. The equation is formulated in equation 6-3, where a is the general coefficient; b, c, d, are the coefficients of effective population, distance, and distance to the next nearest obsidian source, respectively.

$$Q = a + bP + CX + dY \qquad [6-3]$$

Each "source-specific" matrix was analyzed by multi-linear regression analysis, using the computer program, BMDP1R (Jackson and Douglas 1975). The results are presented in table 6-2.

Even though 90% of the exchange systems are not circular in form and asymmetrical, the general linear model holds for 70% of the sources. The values of multiple-R, presented in table 6-2, indicate that 68% to 88% of the variability of the data for the 10 systems is explained by the three variables and the linear model. The notable exceptions are the Fish Springs, Clear Lake, and St. Helena systems. It is not understood why these systems deviate from the model. Among the three variables, distance has the greatest power of prediction shown by the correlation Q-X. Further inspection of table 6-2 indicates that effective population was less significant than distance. However, there is a high degree of correlation between distance and population, correlation P-X. This result may be due to errors of the original population estimates. The surprising result is that the variable, distance to the next nearest obsidian source, appears to be a less significant variable in this

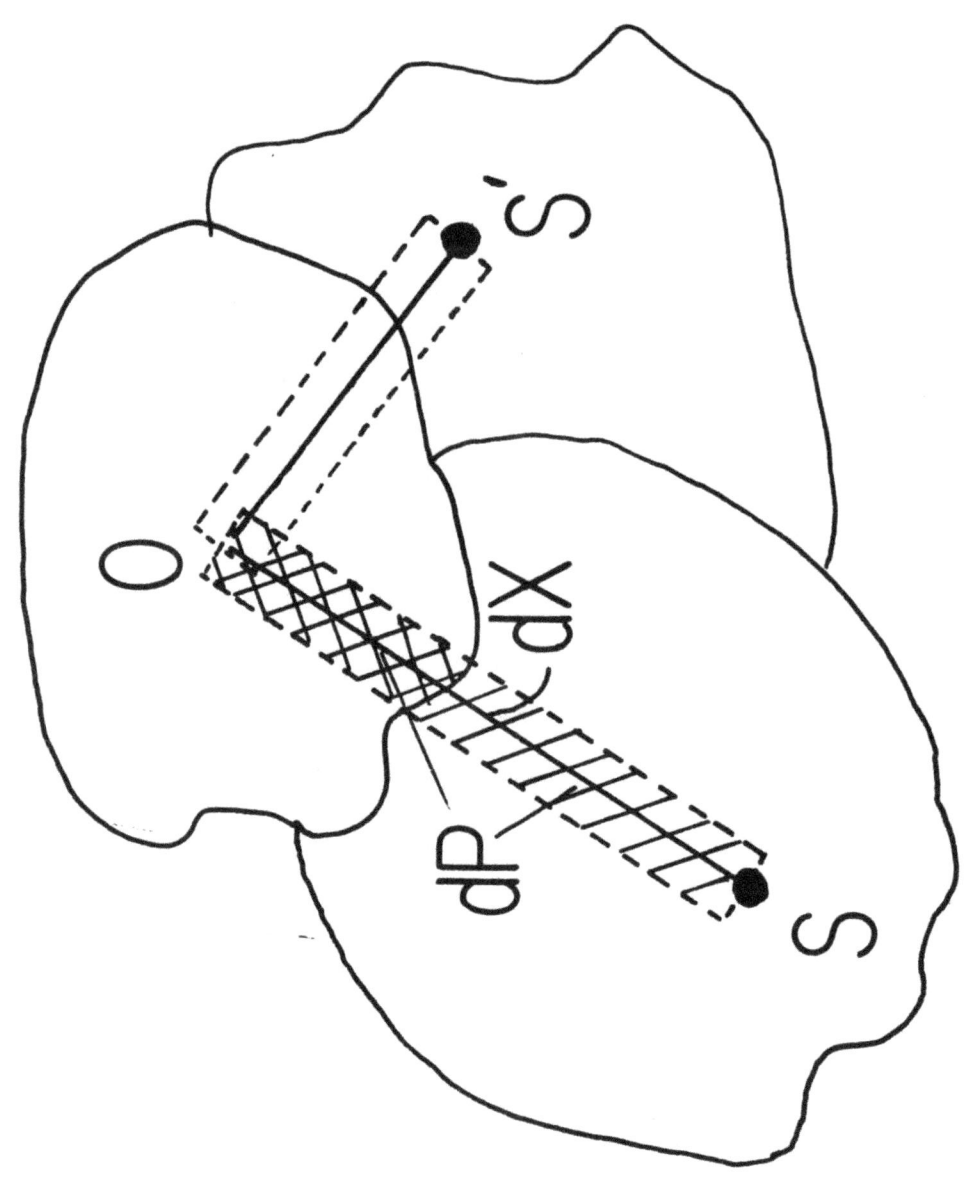

Plate 6-13. Measurement of the Effective Population and Distance from a Source to a Point of Consumption. Two obsidian sources, S and S´, from point of consumption, O, where dP is the partial population contained in each ethnolinguistic segment and dX is the distance between the point O and source S. The width of the segment is 10 miles.

Table 6-2

RESULTS OF MULTILINEAR REGRESSION ANALYSIS OF THREE SYSTEMIC VARIABLES OF THE TEN EXCHANGE SYSTEMS

	Medicine Lake	Obsidian Butte	Bodie Hills	Casa Diablo	Surprise Valley	Annadel	Fish Springs	Coso	Clear Lake	St. Helena
Multiple	0.8659	0.9385	0.7359	0.8337	0.8895	0.8305	0.5386	0.8512	0.6597	0.4395
Degrees of freedom	3/18	2/19	3/24	3/32	3/20	3/8	3/9	3/18	3/21	3/25
F ratio	17.977	70.157	9.452	24.308	25.254	5.928	1.226	15.777	5.395	1.931
P (tail)	0.00001	0.0000	0.00026	0.00000	0.00000	0.01976	0.35575	0.00003	0.00653	0.14051
Sample size	22	22	28	36	24	12	13	22	25	29
Correlation										
$Q-P$	−0.3780	−0.7130	−0.5972	−0.4083	−0.7993	−0.6387	−0.4142	−0.5035	−0.6287	−0.2069
$Q-X$	−0.7561	−0.9363	−0.6737	−0.7655	−0.8433	−0.8290	−0.2717	−0.8361	−0.6447	−0.0371
$Q-Y$	−0.1893	nd	0.4159	−0.1098	−0.2885	−0.8182	−0.2585	−0.3333	−0.4471	0.1894
$P-X$	−0.7887	0.8020	0.8698	0.7132	0.7132	0.7290	0.0635	0.6511	0.8689	0.8837
Coefficients										
a	57.9321	82.3712	40.7956	77.7533	64.3954	32.8587	18.6332	64.3318	49.9648	29.2515
b	0.014	0.008	0.002	0.001	−0.023	−0.001	−0.007	0.001	−0.004	−0.009
c	−0.929	−0.789	−0.388	−0.649	−0.348	−0.386	−0.146	−0.667	−0.324	0.122
d	0.396	nd	0.810	0.346	−0.016	0.029	0.140	0.089	0.037	0.367

Table 6-3

THE CHARACTERISTICS OF THE TEN EGALITARIAN EXCHANGE
SYSTEMS IN CALIFORNIA AT Q = 10%, BASED ON
STEPWISE REGRESSION ANALYSIS

	Radius Catchment (km)	Estimated Total Population
Annadel	98	30,000
St. Helena	206	392,000
Borax Lake	138	110,000
Medicine Lake	209	31,000
Surprise Valley	241	36,000
Bodie Hills	226	98,000
Casa Diablo	256	354,000
Fish Springs	68	17,000
Coso	162	74,000
Obsidian Butte	159	21,000

analysis except for the Annadel system. This is surprising if one considers that the analysis in chapter 5 indicated the effects of alternative sources of lithic materials like chert. The coefficients in table 6-2 provide a linear model of each system following equation 6-3.

Finally, the catchment or area of each system, the boundary of which is limited to $Q = 10\%$, and estimated total population or potential number of consumers within each system was calculated using stepwise multiple regression analysis (Jackson 1975) of the above data. The stepwise regression determined three regression equations. The simultaneous solution of these three equations, setting $Q = 10\%$, provided a means to estimate the radius of catchment and estimated total population which are presented in table 6-3.

CONCLUSIONS

An understanding of regional exchange systems is definitely increased by comparative systems analysis where the difference of structure, internal organization, and continuity can be linked to particular systemic variables. In this study, the isolation of 10 overlapping regional exchange systems has been accomplished by both spatial analysis and chemical characterization of artifacts. The differences of the morphological properties of each system suggest that certain internal parameters may be quite different.

The significance of distance as a measure and variable is not clearly understood. It is suggested that the significance of distance as a systemic variable depends upon the mode of acquisition of specific items and their local importance. A further examination of its significance suggests that the number of consumers may also enter in as a variable.

A general multiple linear model explains 44% to 88% of the variability of the exchange data for 70% of the systems examined. The variables (distance, effective population or potential number of consumers, and distance to next nearest obsidian source) are evaluated by multi-linear regression analysis for each of the 10 exchange systems. Although distance is the best predictor variable, the effect of population or an equivalent archaeological measurement on diminishing supply within a system should not be neglected in subsequent analysis. Even though the location of alternative lithic (non-obsidian) materials is a definite factor which influences the morphology of the systems, the distance to the next nearest obsidian source appears to be a less significant variable in this study. Perhaps the distance to the next nearest lithic source should be considered. Two properties of each system, such as the radius of catchment and estimated total population, allow a direct comparison of the systems.

In closing, California offers an ideal setting in which to study the development and organization of exchange systems. It is here that the high diversity of the environment, which creates a mosaic of localized raw materials and biological communities, sets the necessary preconditions for exchange. In the future this diversity will enable us to evaluate the relative importance of a number of systemic variables. In this way, the conditions and factors which underlie the development and stability of exchange systems among hunter-gatherer subsistence economies will be isolated.

Chapter 7

EXCHANGE SYSTEM GROWTH AND CHANGE

INTRODUCTION

The three previous chapters have developed a methodology for describing regional exchange systems and then evaluating a number of systemic variables. These results suggest a number of important factors which underlie the formation of the systems. The systems, in responding to these factors, appear to be almost mechanical entities organized to gain space utility over specific resources. It has been argued in chapter 4 that exchange systems reduce the amount of work involved in gaining resources when compared to direct access. Furthermore, it has been suggested that there may be a threshold value, dependent on the population, where exchange systems will spontaneously develop. It is of particular interest that this study has shown that source-specific systems such as obsidian and alternative lithic sources interact with each other. These observations of systemic interaction suggest that a model which integrates the formation and mechanism of stability to the respective systems would be very useful. This work would not be complete without suggesting some of the possible mechanisms of systemic change and internal stability. Likewise, a demonstration of the methodology which documents the diachronic interaction of several systems is useful in studying the hypotheses presented.

MECHANISM OF FORMATION, STABILITY, AND CHANGE OF EXCHANGE SYSTEMS

Although a careful consideration of the basis of regional exchange and mechanisms of formation, stability, and change of systems is fully beyond the scope of this work, this study has produced a number of underlying patterns which, upon closer examination, may provide some rudimentary information for further investigation. First, it is important to consider some factors which might stimulate or retard the formation of regional exchange systems.

One factor which would tend to impede the growth, i.e., extent of a system and the amount of goods exchanged, of regional exchange systems is group mobility. High group mobility would favor transportation of small, durable items of diverse materials. Although high group mobility, e.g., year-round, would increase the potential of interaction at many junctures with adjoining groups, the predictability of obtaining critical amounts of specific items, based upon direct contact, would be very low. The lack of scheduling and accounting due to high group mobility are seen as limiting factors which greatly impede the <u>reliable and continuous</u> flow of goods and

and services, and in turn, the formation of stable, regional exchange systems. In this case, direct access by individual expeditions (Webb 1975:366), silent trade (Price 1967), and ad hoc ceremonialism (Vayda 1967) would provide alternative means to acquire particular items from the surrounding region. While high group mobility is maintained, the groups most likely are highly self-sufficient. It is apparent that the growth of exchange systems most likely requires an increased dependency on regional resources and a certain level of interaction of the segments of the population.

As discussed in chapter 4, population growth may be an autocatalytic factor which promotes (and is promoted by) the development of exchange systems. In addition, sedentism coupled with increasing population growth perhaps are the necessary preconditions for the formation of regional exchange systems. With increasing sedentism (which changes man-land relationships), scheduling and accounting improve for groups, yet at the expense of restricting their resource bases. Thus it is expected that the formation of a system is dependent to a degree on sedentism, population growth, and a growing dependency of the population on regional resources.

As it has been previously argued, the formation of a system is most likely a stepwise process resulting from a mixed strategy of direct access and rudimentary exchange relationships. The areas between critical resources have the greatest potential for initial system development with other surrounding areas participating at a later point in time. This stepwise formation process and the effect of the importance of critical resources would tend to produce a bifurcated or branching network of trails. As more items and people are recruited into the system of interaction, the main trunks of the network begin to carry a large diversity of types of items through the process of linking more resources. Secondary and tertiary bifurcations would link the locations of other secondary resources or areas of consumption. The system, and particularly the network of trails, is directly influenced by the spatial arrangement of these individual resources which are recruited into the regional system. It would make a difference whether individual resources were procured at random within a zone, e.g., the procuring of albino deer, or dispersed within a zone, such as the presence of flint outcrops within a sedimentary deposit, or concentrated at a point as are most obsidian sources. Too, the location of alternative sources for potentially replacing each exchange item would affect the system: for example, the occurrence of chert would affect an obsidian-shell trade. It is apparent that the system would undergo many forms of adjustments as it developed. For purposes of analogy, two highly efficient natural network systems are usable as models for exchange system formation: the central nervous system and a river drainage system (Krumbein and Graybill 1965). Both systems have ranked bifurcated channels. The position of each bifurcation is stationary and energetically determined. The rank of a bifurcation is proportional to its expected (or normal) load. Areas within the system which have higher expected loads have proportionally more bifurcations. Both have evolved out of natural conditions in response to ongoing processes and functions they serve. It is hypothesized that a regional exchange system evolves as a natural process into a system having a ranked, bifurcated network of trails in a manner analogous to the other systems developed for increasing the efficiency of distribution of resources.

A closer examination must be given to the regional resources (cf. Wright 1974). Regional resources can be divided into an ordinal scale: (1) critical or <u>vital resources,</u> materials which cannot be substituted for in the performance of necessary subsistence activities of a population and are consumed generally by members of that population, e.g., salt; (2) utilitarian (Tourtellot and Sabloff 1972) or <u>superior resources</u>, materials which can be necessary for subsistence, but represent a material which is superior in quality as compared to the other available resources. These may be dietary supplements or superior raw materials for technological items; however, these may or may not be shared by all members of a given group, e.g., obsidian vs. quartzite, digger vs. sugar pine nuts; and (3) <u>luxury resources</u>, materials which are usually rare and act as sociotechnic items (Binford 1965) for individuals or the group, e.g., a shell product. The function and definition of a particular resource is dependent on the group and local resource base. For example, if the regional geology contained only granite, pure mature sandstone, or pure limestone, obsidian or chalcedony would become a <u>vital</u> resource for flake-tool technology. With the addition of non-granite roof-rocks or conglomerate, the people of an adjoining region would probably consider obsidian/chert as <u>superior</u> resources for their technology. In yet another region with abundant chert, obsidian, a scarce resource, might be considered a <u>luxury</u> resource used only for status items. This example suggests the differential use of materials dependent on their scarcity. Thus it is suggested that the relative ratios of the three resource categories, due to differential consumption or quantity of an item, may follow the relationship: $m/_D X :: n/P :: P$, respectively, luxury, superior, and vital resources where P is population, m and n are variables related to work expenditure, and X is some exponent of D, distance (cf. Renfrew 1977). As already argued in chapter 6, vital resources will be independent of source distance within that area within the system, and their importance would be dependent on the total arrangement of similar resources. The total amount of vital resources consumed within that area of the system is a function of a population, i.e., recovery of durable vital resources may be used as an index of prehistoric population. The change of value of all items from juncture to juncture underwrites a certain amount of consumption at each juncture (Wright and Zeder 1977). As such, by changing the value, the middle group can remove items out of the system for their personal consumption without adding anything to the exchange item inventory which is illustrated in plate 7-1. However, in the face of a declining value system, certain middle groups may have to add items to the inventory. If such a group does not have a local resource, it may be forced to increase the value of an item by altering it into a usable product.

The formation of a system does not automatically insure its stability and maintenance. Stability is gained by a number of mechanisms. This may be an improvement of the scheduling and dependability of acquiring regional resources. These factors are the following: (1) a growing population moving toward increased sedentism, (2) the gradual linking or coupling of different classes of critical resources as binary pairs, (3) the recruitment and modulation resulting from a wide diversity of items both superior and luxury (cf. Wright and Zeder 1977), and (4) the gradual growth of a bifurcated network of ranked trails. These internal modifications of system formation are complex, localized, and difficult to predict, except that on the whole, they

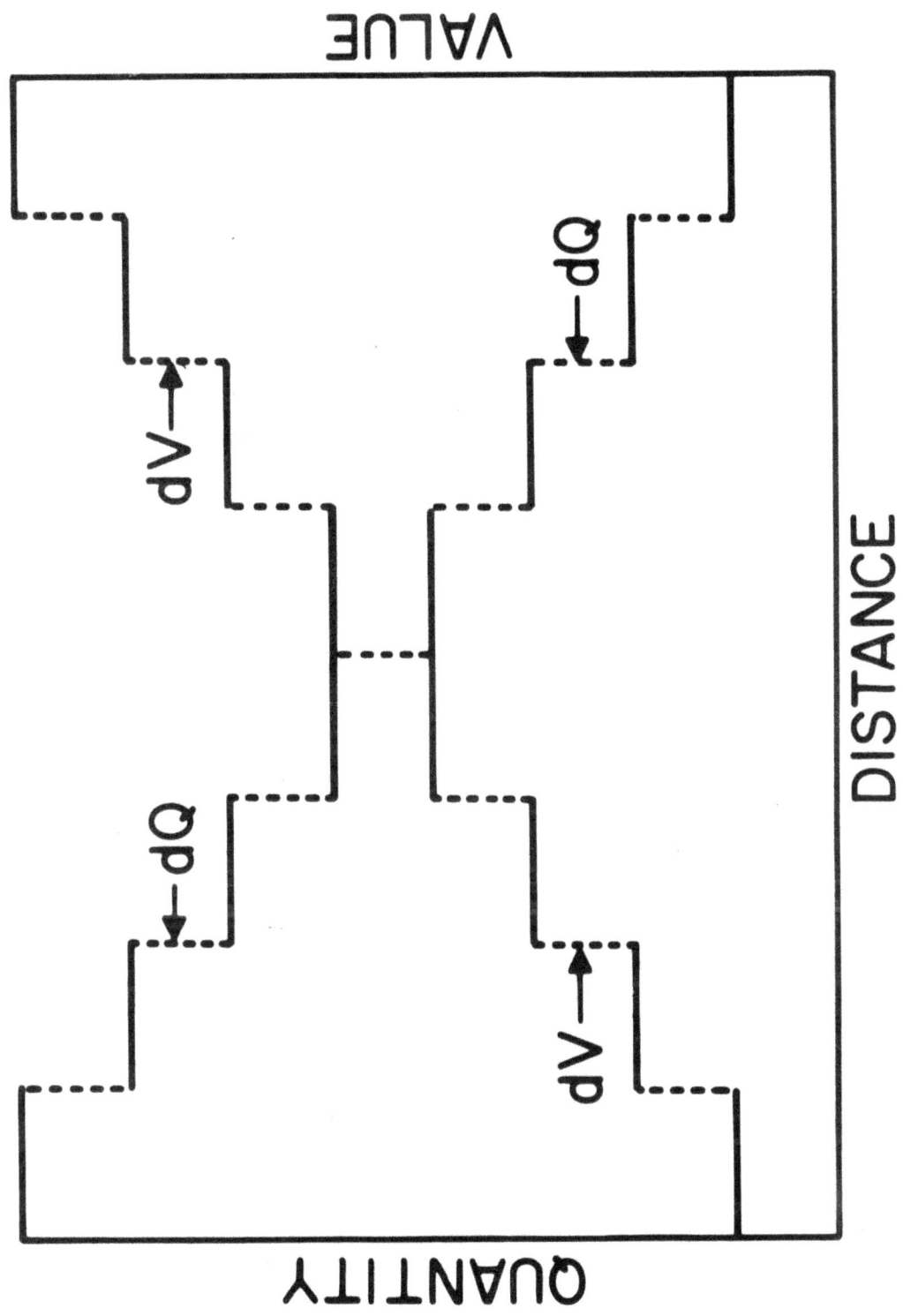

Plate 7-1. A Hypothetical Case of How the Increased Evaluation of a Material Can Support its Consumption within an Exchange System.

appear to work. Eventually it will be possible to show that systems change toward increasing their efficiency in supplying a diversity of regional resources.

As a thermodynamic rule, the more highly organized a structure, the more inflexible it is. To gain flexibility, valuables and primitive money assume an importance, as pointed out by Rappaport (1968) and Wright and Zeder (1977). The functions of primitive money are at least the following: (1) it is exchangeable for most items, (2) it gives tremendous flexibility to members in any transaction, and (3) to the system as a whole. Wright and Zeder (1977) suggest that primitive money underwrites the distribution of vital resources. The predetermined value of items within an exchange system alludes to two underlying features of the system: (1) a quid-for-quo value of items at a particular location implies a certain expectancy of receiving items as ratios of each other, and (2) as forementioned, the change in value of items within a system actually can underwrite a certain amount of consumption at each juncture (cf. plate 7-1).

Whereas it has been argued that the division of the vital resource expects as many divisions as consumers, luxury resources may be independent of population. It is expected that the physical form of primitive money and luxury resources would tend to reduce any further modification once entered into the exchange system, e.g., deer skins and shell bead strings. In fact, deliberate burial or change in the stylistic elements of primitive money would counter its inflation, because supply would not be permitted to expand indefinitely. In the previous discussion, it has been maintained that the primary function of an exchange system has been to homogeneously distribute its vital resources, formerly acquired by direct access. However, considering concomitant changes in population, increased demand would have put the primary production of vital resources under stress. If it is assumed that the demand will be met, several adjustments could be made to compensate for the increased demand for these resources. The obvious response would be to increase the number of producers at the source point as suggested by Wright and Zeder (1977). This could be accomplished by several means, namely: (1) increase the population by removing birth control sanctions, if they existed, (2) incorporate more producers by relinquishing exclusive rights to the source, if they existed, and reestablishing local direct access to surrounding groups, and (3) underwrite the subsistence needs of local specialists, who were formerly involved in primary subsistence tasks. Another alternative response would be to change the technology of production, namely: (1) to spend less time or effort on producing each item, (2) to change from specialized products (finished hand axes) to more generalized products (flakes or preforms in a variety of sizes), and (3) to change the procedure or steps of production making it more efficient. It is obvious from the above that the production centers or point of input of vital resources are critical in understanding regional interaction, simply because these centers are sensitive to systemic changes. These changes may take the form of being short-term random or cyclic oscillations or long-term trends within the system or abrupt and radical discontinuities.

In general, systemic change is difficult to predict. As a phenomenon, it has been a popular topic of discussion throughout the social sciences: a critical examination here is beyond the scope of this work. Nevertheless, in the sections which follow, it will be shown that the systems under investigation are not static, though stable in their operation. Systemic change is the result of one or more systemic variables which have been discussed above. For example, if the internal structural components do not stablize the system or if the production centers cannot respond to increased supply or demand, or if an extensive social (population movement) or technological change occurs, then these may, in fact, promote systemic change. It is expected that the production centers are the best points to monitor variation or detect change within the system. For that reason, the next section documents the diachronic production rates of three interrelated systems.

DIACHRONIC PRODUCTION RATES OF THREE INTERRELATED EXCHANGE SYSTEMS

In chapters 5 and 6, a methodology was employed which described the synchronic source-specific regional exchange systems based on obsidian consumption. This same methodology could be adopted to study these same systems in a diachronic perspective. However, in California, with the need to distinguish the many sources by chemical characterization, the cost would be prohibitive as discussed in chapter 1. This would be compounded by the expense of radiocarbon dating to insure adequate chronological control. It is here that the application of obsidian hydration dating is demonstrated to be most useful for gaining chronological control in describing diachronic production rates at three obsidian production centers.

The three major exchange systems in Central California, the St. Helena, Bodie Hills, and Casa Diablo systems, will be examined over a period of at least 1,500 years. The St. Helena exchange system will be examined through data obtained from three neighboring sites, CCo-30, CCo-309, and CCo-308 studied by Frederickson (1969). The Bodie Hills system will be reexamined through production analyses at the quarry Mno-612, performed by Singer and Ericson (1977). The Casa Diablo exchange system is viewed from the village-quarry workshop at the Mammouth Junction site, Mno-382 (Michels 1965, n.d.; Sterud 1965). From each site or set of sites the production rate curve is monitored using radiocarbon or obsidian hydration dating to gain chronological control over the data.

The St. Helena System Viewed from CCo-30, CCo-309, and CCo-308

The St. Helena exchange system with the obsidian source in Napa County will be observed from three sites which are 85 to 90 km to the southwest near the modern town of Alamo in Contra Costa County. There may be an inherent weakness to view the production from afar. However, these sites provide an almost continuous record of the use of specific lithic materials for chipped-stone-tool production beginning approximately 2000 B.C. to Late Horizon times in the eighteenth century, as a result of extensive research (Curtis and Frederickson 1964; Frederickson 1965, 1966, 1968, 1969; Jackson 1974). Although this research provides the necessary controlled data for a diachronic study, quarry production analysis of the St. Helena source would provide the best and most comprehensive picture of this important obsidian exchange system. (Research here should be given high priority now that its importance is more clearly understood. However, a project is hampered by a bountiful summer crop of poison oak.) A further limitation of the data is that the data represents consumption rather than production. Finally, these three sites form only a single node within a much larger and complex exchange system. Most likely the rate of consumption of St. Helena obsidian is quite different if viewed from other points in space. Nevertheless, it does appear that a general diachronic view of the St. Helena exchange system can be obtained from data derived from these sites.

The three sites do provide a nearly continuous record of the utilization of lithic materials for the last four millennia. Access to this area from the north requires crossing by boat at the Carquinez Straits or traveling on an overland route through the Sacramento River delta area to the east. The intensive study by Frederickson (1969) indicated that there was evidence for the "immergence of trade" in this area in the Late Horizon. Since that study obsidian artifacts and debitage have been chemically characterized by X-ray fluorescence analysis (Jackson 1974) and neutron activation analysis presented in appendix 2. Both of these studies indicated that nearly 100% of the obsidian originated from St. Helena or Napa Glass Mountain. In addition, a total of 13 new radiocarbon dates were run on samples from these sites to gain additional chronological control in conjunction with formulating the hydration rate for this obsidian source. It is important to present some basic information on each site; this information is summarized from Frederickson (1969).

The Rossmoor site (CCo-309) located 4 km north of the town of Alamo, has a single Phase 2, Late Horizon cultural component, representing a 200-year occupation (1500-1700 A.D.). A charcoal sample associated with a Phase 2 cremation within the midden was dated at 285 ± 95 years B.P. (I-1193). The percentage of obsidian utilized relative to chert, the index devised by Frederickson, is 92.4%. The on-site lithic material is sandstone, although chert and a variety of other materials occur within the midden; constituent analysis revealed that there was a greater utilization of molluscan food resources than game animals, relative to earlier cultural components in this area (Curtis and Frederickson 1974; Frederickson 1969:113-115).

The La Serena site (CCo-30) located 0.8 km south ($37°50'$ N. Lat., $122°1'$ W. Long.) of the modern town of Alamo contained two cultural strata separated by culturally sterile soil. The earlier component as attributable to the Middle Horizon indicated a 0.8% obsidian utilization at this time. The later cultural component is attributable to Phase 1 of the Late Horizon with intensive use occurring between 750 and 1500 A.D. A series of four radiocarbon dates (UCLA 1793 A-D) on charcoal excavated from different unit-level proveniences, verified their late Phase 1 cultural affiliation having a range of 1380-1460 A.D. The percentage of obsidian utilized was 56.2%. Numerous lithic materials such as chert, quartzite, and sandstone were available in San Ramon Creek which runs past the site. The variety and dominant species of shellfish are very similar to those exploited in CCo-309 (Frederickson 1965, 1969: 106-112).

The Stone Valley site (CCo-308), located in the town of Alamo ($37°51'$ N. Lat., $122°1'$ W. Long.) is the earliest of the sites under consideration. Three distinct cultural components were identified within physically discrete strata to a depth of over 20 feet. The uppermost cultural stratum (Component A) was assigned to the Middle Horizon/Late Horizon transitional phase, dated to approximately 100 B.C. to 300 A.D. The middle component (Component B) was assigned to a middle Middle Horizon phase, "guess-dated" circa 700 B.C. The third and deepest component (Component C) was assigned to an early Middle Horizon phase based upon material affiliations and a radiocarbon date of 4450 ± years B.P. (UCLA-259) on excavated charcoal. Nine new radiocarbon dates were measured on charcoal derived from the three components (UCLA 1786 A-C and 1792 A-F). They suggest an extensive cultural lag (750-1050 A.D.) in the beginning of the Middle Horizon/Late Horizon transitional phase which has been dated at 100 B.C.-300 A.D. in other areas in Central California (Frederickson 1966, 1969:115-116). The percentages of obsidian used during Components A-C are respectively, 6.9%, 2.4%, and 1.8% (Frederickson 1969:Table 8). The time period of Component A is revised to 750 to 1050 A.D. based on radiocarbon dating. In this study, the earliest period in interior Contra Costa County was a time of dependence on local game and lithic materials and lacked well-developed exchange relations with areas from which molluscan foods and obsidian were derived (Frederickson 1969:120.)

In his original study, Frederickson presented a chronological comparison of proportional midden constituent indices (Frederickson 1969:Table 8). Now with the subsequent research of Jackson (1974) and the data presented herein, it has been determined that the obsidian was originally derived from the St. Helena source during this time based upon a chronological sequence developed by 15 unpublished radiocarbon dates. The only essential revision is the time of the Middle Horizon/Late Horizon transition phase, which has been revised to 750 to 1050 A.D. in this area.

The diachronic rate of obsidian consumption as seen from interior Contra Costa County is shown on plate 7-2. As pointed out earlier, this curve shows in general that the rate of exchange of obsidian increased rapidly beginning at the Middle Horizon/Late Horizon transitional phase, circa 750 to 1050 A.D. The rate of exchange increased continuously until the Protohistoric period. It is interesting to note that obsidian tools in the earlier periods, in whatever quality they were present, were probably imported as complete tools (Frederickson 1969:120). However, in the Late Horizon the local manufacture of tools takes place on-site from imported raw materials (Frederickson 1969:123).

Diachronic Production Rate at Bodie Hills Obsidian Quarry

Singer and Ericson (1977) present a complete and careful study of the diachronic production rate at the Bodie Hills obsidian source, which is the northeastern source. It is important here merely to summarize some of the basic findings which have a direct bearing on the interrelationship of the three sources under investigation.

The production analysis which was conducted at Bodie Hills focuses on both the lithic technology used at the quarry and a quantitative analysis of the diachronic rate of production. Obsidian hydration dating is used to determine the time of use of specific quarry workshops, shown in plate 7-3,

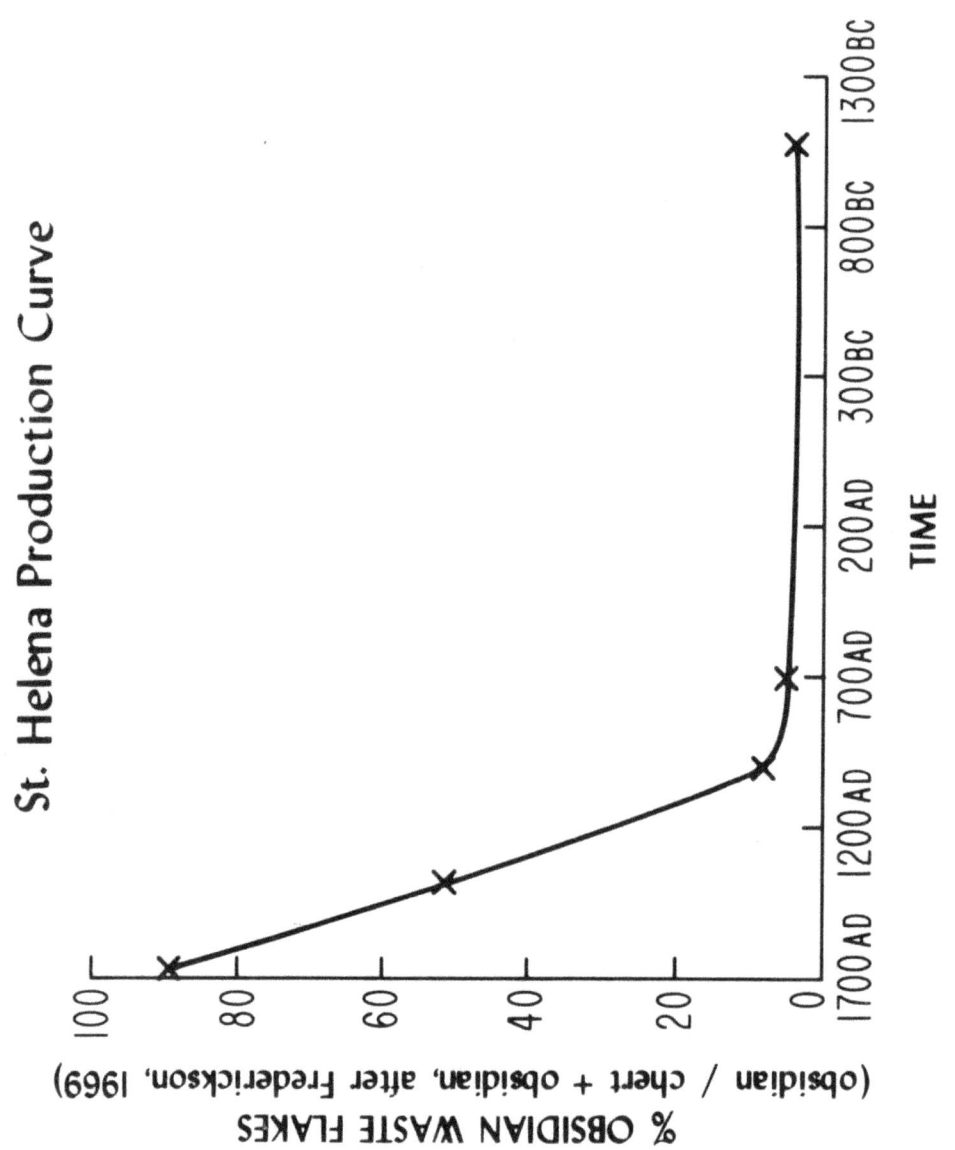

Plate 7-2. St. Helena Exchange System Obsidian Production Rate Curve. This diachronic production curve was drawn from combining the data information derived from the obsidian and chert utilization indices presented by Frederickson (1969) and associated radiocarbon dates (Appendix 2) from three Contra Costa archaeological sites.

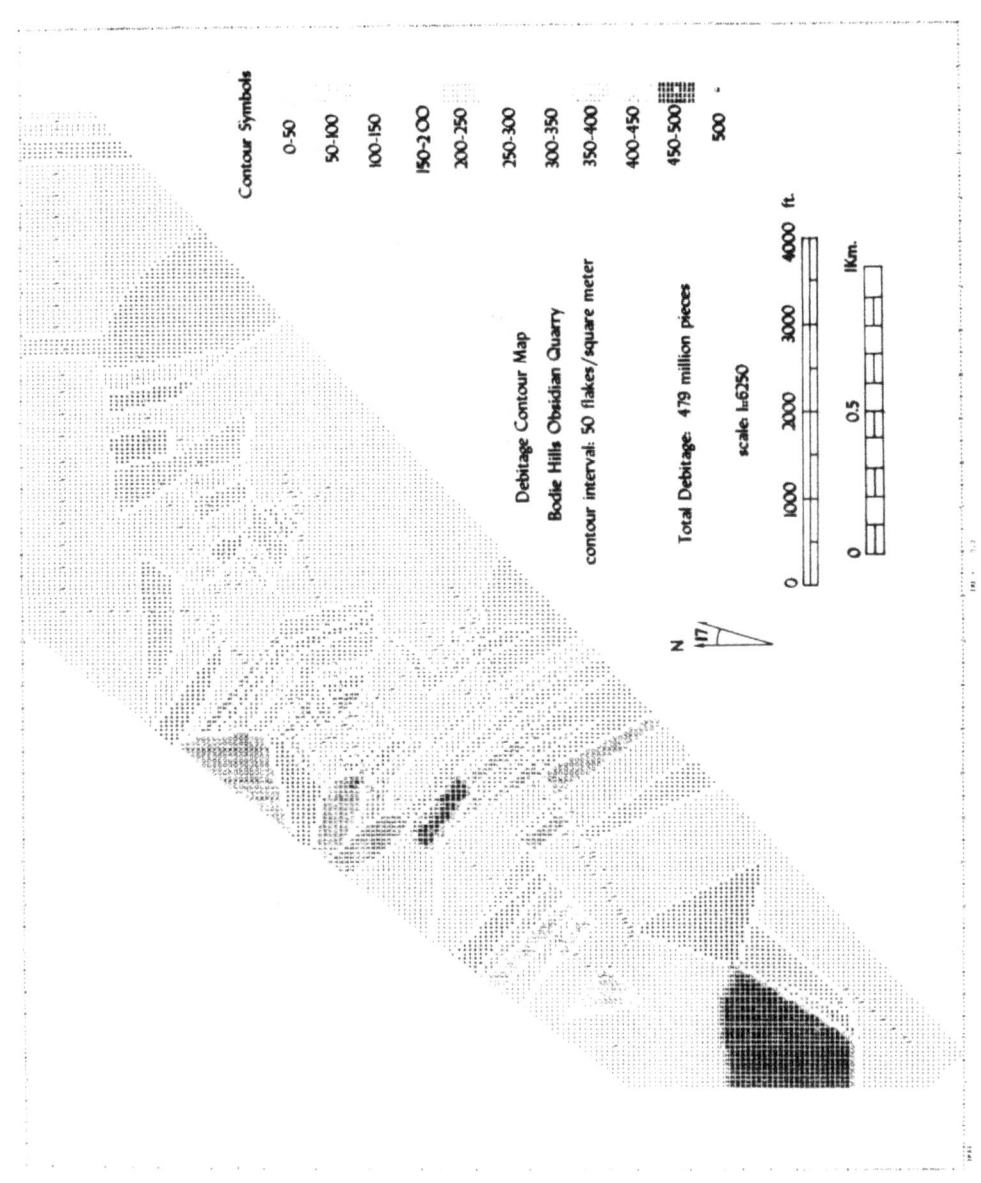

Plate 7-3. Synagraphic Map of the Debitage at the Bodie Hills Obsidian Source, California (after Singer and Ericson 1977), Bodie Hills Exchange System, Obsidian Production Rate Curve. The curve was drawn using data presented by Singer and Ericson (1977).

indicated by synagraphic mapping of intensive surface data collected along random transects.

The diachronic production rate curve shown in plate 7-4 was derived by plotting the frequency of hydration measurements and their respective ages using a linear obsidian hydration rate of 650 years/micron. As a result of improved data, this value has changed to 670 years/micron (see table 1-12). The former value is used here to avoid recalibrating the earlier study. (The correction would be minimal.)

The Bodie Hills production curve has some very interesting features shown in plate 7-4. Production began approximately 5,000 years ago, synchronous with the Early Horizon of Central California. Heizer (1974) noted that Bodie Hills obsidian was used in making Windmiller culture artifacts. A relatively constant production rate is maintained until approximately 2,000 B.C., which is synchronous with the beginning of the Middle Horizon (Heizer 1964:127). The production rate rapidly increases at this time, reaches a maximum at approximately 1,000 B.C., and then diminishes and almost terminates with the close of the Middle Horizon dated to approximately 500 A.D. (Singer and Ericson 1977:186). The reasons for this abrupt change of production of Bodie Hills obsidian are now more apparent in light of the systems under investigation. What is critical here is observing the synchronous changes in lithic technology. There are two modes of lithic production represented at the Bodie Hills site illustrated in plate 7-5: an early, large biface production mode linked to "luxury" exchange where the bifaces appear in burial context (Ragir 1972); later a generalized blade production mode related to utilitarian exchange appears as the context of obsidian utilization becomes very generalized.

Beginning with the earliest production at Bodie Hills, large bifaces were produced. There was evidence of their early production in several workshop areas, particularly in the Macro areas having a maximum exploitation between 2250 and 5000 B.P. (Singer and Ericson 1977:183). In time, several biface forms occur, but gradually there is a reduction in their size throughout the Middle Horizon. Although the Bodie Hills site is an immense workshop with an estimated 479 million pieces of debitage produced in 5,000 years of its existence (Singer and Ericson 1977:187), the average annual biface production was estimated to be 960 to 1,725 bifaces a year (Singer and Ericson 1977:185). If these estimates are anywhere near correct, annual production was very limited. In fact, given 98,000 total population, very few people were supplied with obsidian from Bodie Hills, perhaps 1% to 2% per annum. This low estimate is derived by dividing production rate by the effective population presented in table 6-3. This evidence suggests that the Bodie Hills obsidian bifaces were used perhaps as luxury items. It is important to note that biface production is extremely wasteful, creating abundant debitage for each finished piece. As a result, the amount of debitage from biface production greatly exceeds the debitage from blade production, which produces a generalized set of lithic forms for tool production. Compare debitage difference between the two production technologies represented in plate 7-5. If one considers that one finished biface is equivalent to hundreds of small (blade) projectile points, the gradual change in lithic production from finished biface

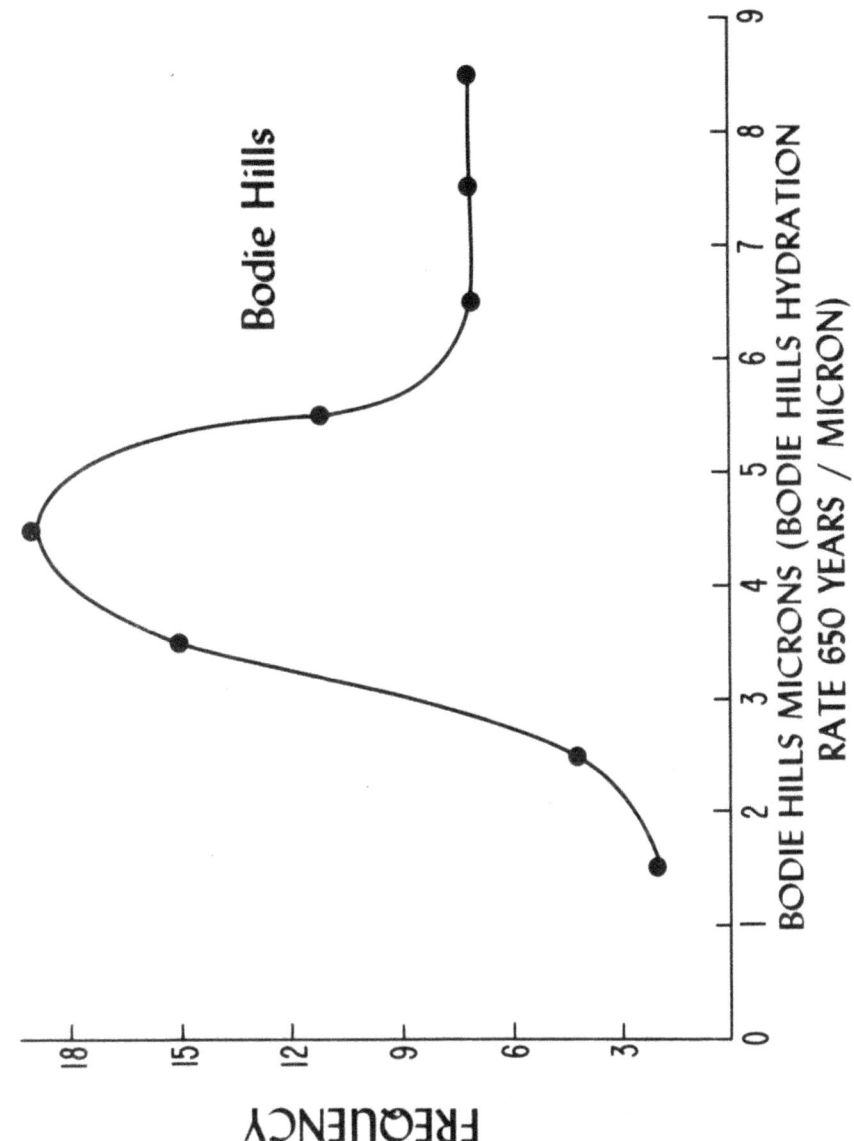

Plate 7-4. Bodie Hills Exchange System, Obsidian Production Rate Curve. The curve was drawn using data presented by Singer and Ericson (1977).

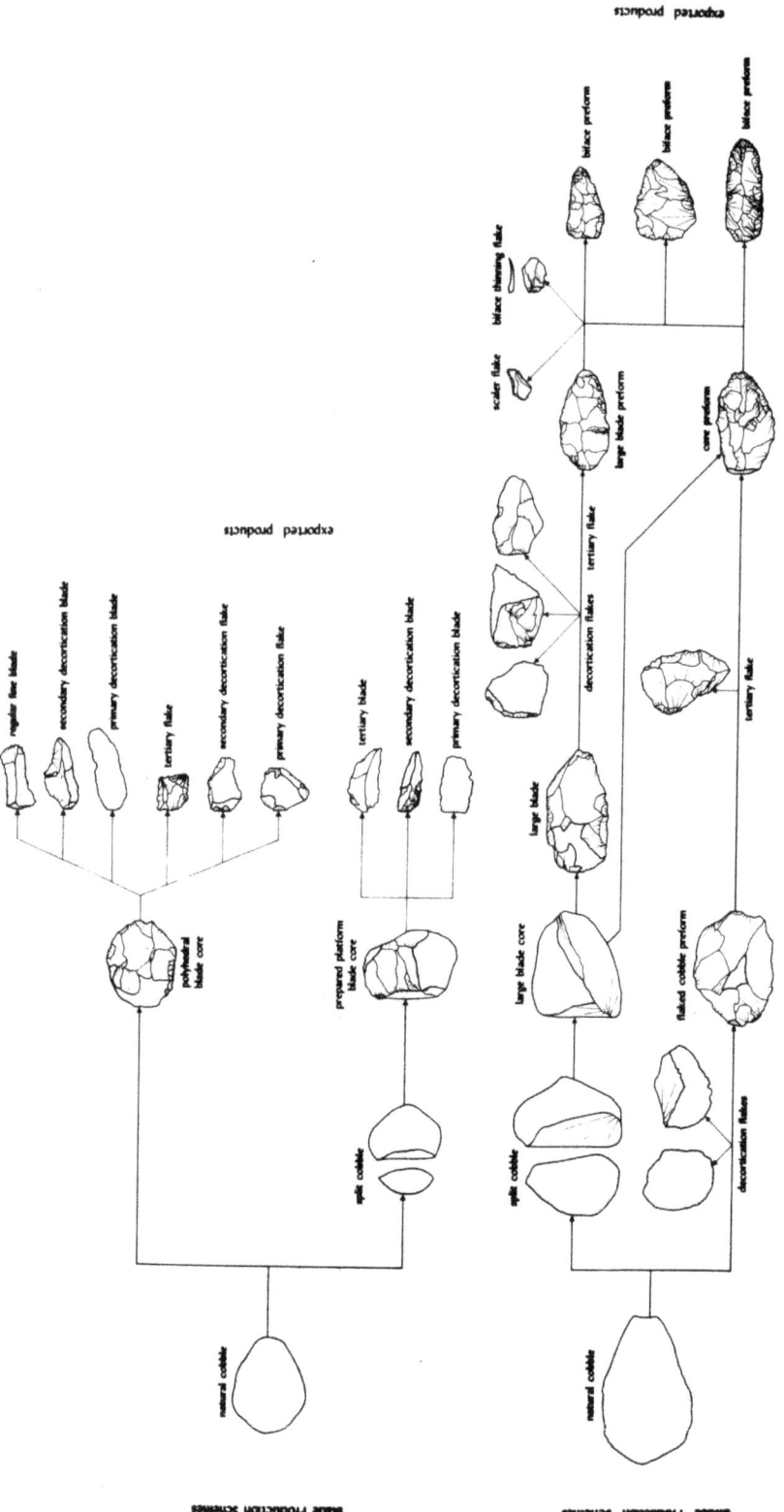

Plate 7-5. Bodie Hills Lithic Production Modes. Discussed by Singer and Ericson (1977). The biface production scheme indicates the forms and amounts of debitage in producing a single biface preform. On the other hand, the blade production scheme indicates the forms and amounts of flakes and blades produced by this lithic industry. A comparison of the two schemes indicates the differences in morphological characteristics and number of exportable products. Actual artifacts were used as models in drawing this scheme.

to generalized blade forms could produce sufficient items to supply an increased population base.

Unfortunately for this study, the Bodie Hills study (Singer and Ericson 1977) does not examine the production of blades. However, it does appear that the workshops of predominantly blade production do occur much later in time, particularly in the Ridge area (Singer and Ericson 1977).

In conclusion, there is preliminary evidence for an early, luxury biface exchange followed by a later, utilitarian blade exchange. In addition, the major changes in the rate of production appear to be synchronous with the "breakpoints" of the Central California cultural chronology. Whether the biface production terminated at 500 A.D., to be replaced by blade production, remains to be determined by additional study. It is obvious that the Bodie Hills exchange system continued to produce obsidian after 500 A.D. as already described in chapter 6. Resolution of these problems is crucial.

Diachronic Production at the Casa Diablo Obsidian Quarry: A View from Mno-382

The diachronic production of obsidian at the Casa Diablo quarry is monitored at the Mammouth Junction site, 4-Mno-382. This site was intensively studied by Michels (1965) for his dissertation. His main thesis was to demonstrate the applicability of obsidian hydration measurements in assessing the problem of stratigraphic mixing. Fortunately, the original data can be utilized here to obtain at least a working diachronic production rate curve for the Casa Diablo exchange system.

The Mammouth Junction site is located immediately adjacent to the Casa Diablo obsidian source. It appears to have been the locus of a number of activities: quarrying, manufacturing, hunting, traveling, and living in summer residence during the course of the last 7,000 years (Michels 1965:39). Four stratigraphic layers were identified and represented in most excavation units (Sterud 1965). A total of 42 5 x 5' units were excavated to average depths of 36 to 48", which yielded 150.4 cubic yards of midden. A total of 9,475 specimens were cataloged under UCLA Accession Number 436. Although most of the artifacts were obsidian, unfortunately for this study, unmodified and unused debitage were discarded during the process of screening (Michels 1965: 33). In contrast to the Bodie Hills quarry where no charcoal or organic materials were observed during the survey, the Mammouth Junction site yielded 20 radiocarbon samples. Six of these samples have been dated. The samples were originally selected for source-specific hydration rate determination. Unfortunately, five of the dates derived from samples at very shallow depths (6" to 12") have a Late Horizon context 1280 to 1660 A.D. As a consequence, the Casa Diablo hydration rate reported in table 1-12 is not very accurate. The projectile point typology chronology developed by Lanning (1963:278), modified by Michels (1965:table 6), and substantiated by Clewlow, Heizer, and Berger (1970) support a Casa Diablo hydration rate of approximately 1,000 years per micron derived by Clark (1964). This rate is used to establish the chronological framework of the quarry production at the Casa Diablo obsidian source. Needless to say, the Casa Diablo obsidian hydration rate definitely needs further work.

Although Michels does not discuss diachronic production at the Mammouth Junction site, he does present a frequency distribution of hydration measurements through time for the total artifact sample (Michels 1965:fig. 40). Michels' frequency diagram is redrawn and the "t-units" are converted to 1,000 years per micron to provide a chronological framework. The diachronic production curve of the Casa Diablo exchange system is presented in plate 7-6. It is immediately apparent that both production curves of Casa Diablo (plate 7-6) and Bodie Hills (plate 7-4) are almost identical in form. The same processes of production and supply must have been operating at the same time.

Of critical importance here are the similarities in lithic production between the two obsidian sources. This comparison is hampered by the lack of an extensive debitage analysis or the availability of debitage for analysis and the differences in artifact typologies used by Michels (1965) and that of Singer and Ericson (1977). In the future, an attempt to correlate the two assemblages and lithic products produced at Bodie Hills and Casa Diablo obsidian sources will be made.

The change in lithic production as a function of time is discussed by Michels in an eight-phase chronological framework. These phases corresponded to give "t-units" (Michels 1965:92) which have been converted in this discussion to calendrical time. Certain observations made by Michels (1965: 197-200) are extremely relevant in this analysis. Heavy bi-pointed knives (bifaces) make their appearance at approximately 5000 B.C. No projectile points or choppers are present at this time. From circa 3500 to 1700 B.C., knife (biface) frequency and type variability increase steadily. Scrapers (blades) also exhibit a steady increase in frequency and type. Choppers are more frequent than in earlier periods. At the beginning of Phase 6 (1600 B.C.- 180 B.C.) there is an abrupt reduction in the projectile point inventory both in type and frequency. This change coincides with the maximum production of knives (bifaces) in frequency and variability. Choppers have the highest frequency now. The productive output diminishes during the end of this phase. From circa 170 B.C. to 300 A.D., the site is abandoned to be reoccupied by people with a "very simple hunting technology." This pattern of increased biface reproduction followed by abandonment is repeated at both Bodie Hills and Casa Diablo during the same period. The production systems appear to "gear up" in response to some stimulus and then gradually fade.

Three Interrelated Exchange Systems: A Discussion

In the preceding pages, changes in products and the rate of production have been examined and synthesized for each of the three systems. Plate 7-7 is a composite diachronic production curve of the three systems. Although the curves are based on different scales, they are comparable. They show that production rates of the Bodie Hills and Casa Diablo systems diminished near the beginning of the Late Horizon of Central California prehistory. The St. Helena exchange system, if correctly viewed from the Alamo area, responds very slowly to the changes in the systems to the east of the Sierras. Perhaps the most significant point here is the "apparent" lack of obsidian exchange from either the western or eastern sources for at least 500 years. This would point to a hiatus in the presence of obsidian in the archaeological record. If this was real, it should have been noticed before now. Obviously, something

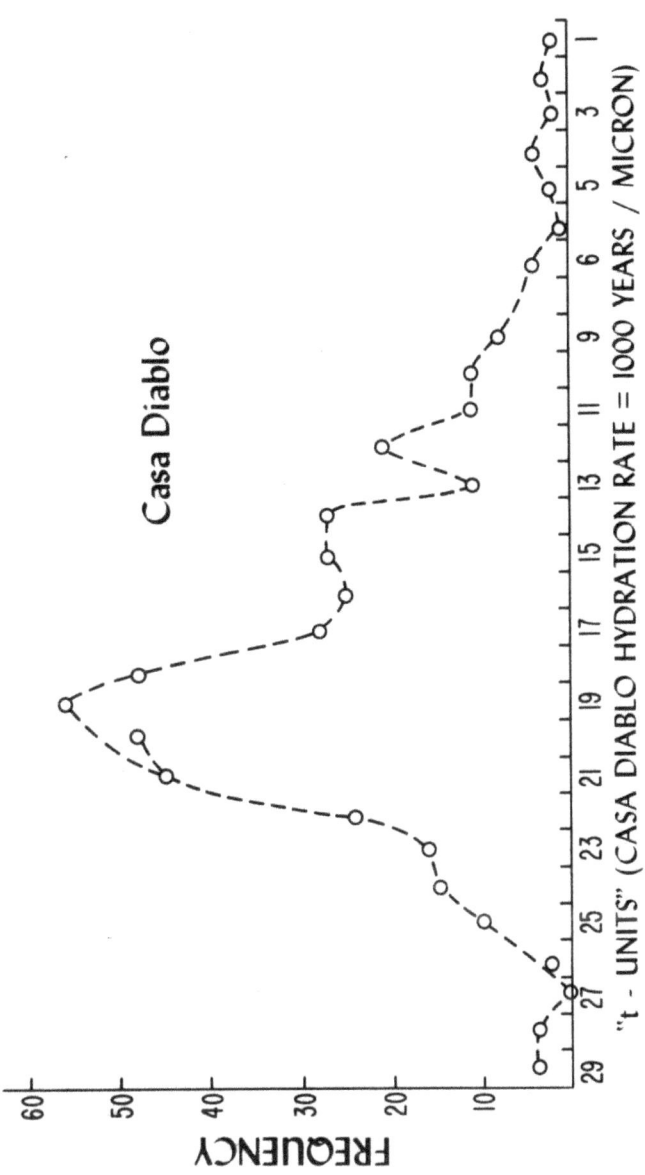

Plate 7-6. Casa Diablo Exchange System, Obsidian Production Rate Curve (Modified after Michels 1965).

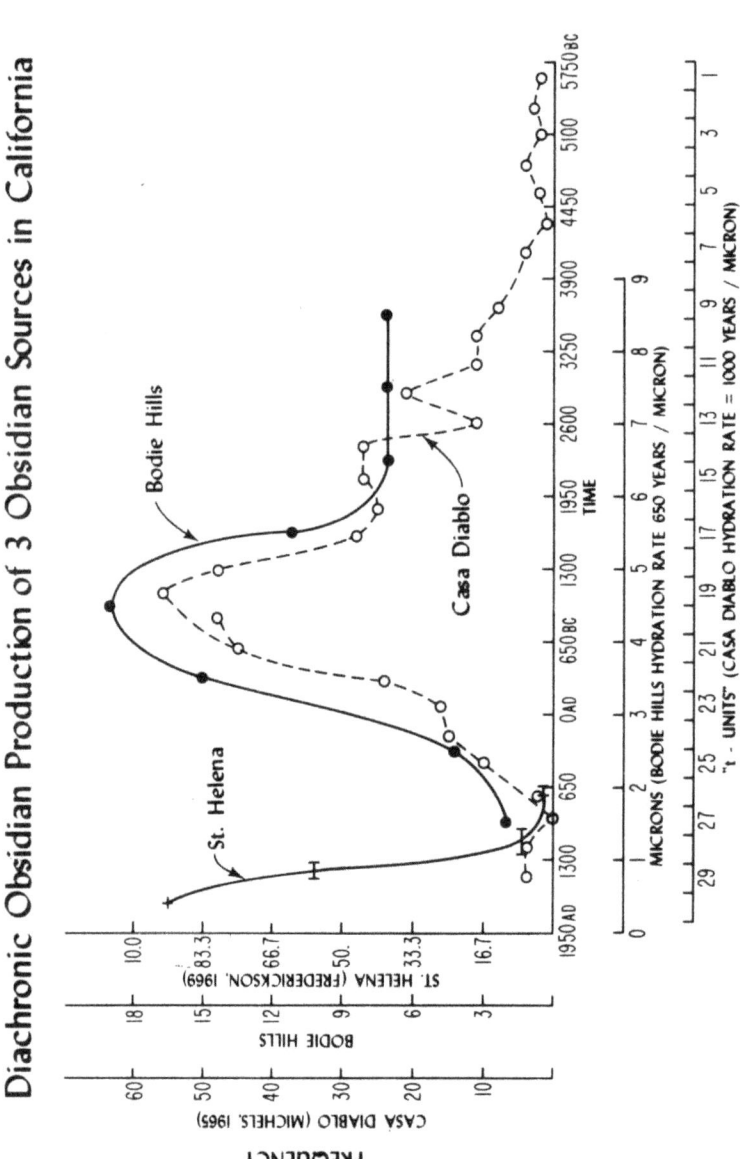

Plate 7-7. Diachronic Production Rates of the Three Exchange Systems under Study. The obsidian production rate curves are grouped in order to understand their interaction. It does appear that there is similarity in the diachronic production rates of the Bodie Hills and Casa Diablo obsidian sources. After the apparent decline in these production rates circa 500 B.C., the St. Helena production rates begin to increase.

is wrong with the analysis presented herein. This obviously stems from analysis of the Alamo area. New evidence can be introduced which resolves this problem.

Jackson (1974) studied obsidian exchange within Central California using X-ray fluorescence analysis. Most of his results tend to supplement the finding of this paper. He does see evidence for an economic shift as viewed through obsidian utilization in the Middle Horizon (1974:78). This change takes the form of a decline of the eastern sources, Bodie Hills and Casa Diablo (1974·70). By the end of the Middle Horizon, Napa Glass Mountain (St. Helena) and Annadel, which remains highly localized as shown in plate 6-5, became the dominant sources in Central California (1974:70). To quote directly:

> Nevertheless, it would seem that sometime between the middle and end of the "Middle Horizon" in this area [referring to the Lower Sacramento and Upper San Joaquin Valleys], Napa Glass Mountain [referring to St. Helena] became an increasingly important source of obsidian. By the beginning of the "Late Horizon," Napa is clearly the source of obsidian for the lower Sacramento and upper San Joaquin groups (Jackson 1974:69).

> While Napa Glass Mountain [St. Helena] is clearly the dominant source throughout the prehistoric record of the Bay Area, there is a parallel decline of the importation of eastern obsidian to that which we see in the Delta and Valley areas (Jackson 1974:70).

This is exactly what was suspected. The Alamo area does not present a representative view of the development of the St. Helena exchange system. The Alamo area must have been a hinterland between the Bay area to the west and the Delta area to the east, i.e., obsidian most likely bypassed the Alamo area. However, in time, the route became more direct. This finding supports the importance of production analysis at the sources rather than at other areas in the system.

Perhaps more important than resolving the above problem, this new information suggests a discernible resource competition between the St. Helena (western) and Trans-Sierran sources (eastern) at the close of the Middle Horizon. These tentative results remain to be verified through analysis of sites in the Delta area. Jackson (1974) discusses some aspects of lithic production which are important. He observes the absence of any lithic debitage from the earlier sites in the lower Sacramento and upper San Joaquin Valleys which suggests that the Windmiller people were receiving completed artifacts. Nevertheless, in later periods the importation of raw material can be documented. Blanks from St. Helena have been recovered and waste flakes become more numerous in Phase 2 of the Late Horizon (Jackson 1974: 79), noted earlier by Frederickson (1969).

Finally, there is evidence to suggest a difference in the lithic technologies between the eastern sources and that of Napa Glass Mt. To quote Jackson directly:

It should be noted, however, that the large exotic blades and points of the Delta Valley area which exhibit the fine "ripple" or "ribbon" flaking are generally made of obsidian from the Bodie Hills (especially the ceremonial side-notched points with elongated blade elements) or Casa Diablo (especially the larger and heavier blades for ceremonial use). Although there are large blades and points manufactured from the Napa Glass Mountain material, these do not demonstrate the fine precision of flaking (Jackson 1974:80).

In summary, quarry production analysis and related information suggest that the eastern sources at Bodie Hills and Casa Diablo both produced bifaces (knives) until about the end of the Middle Horizon, when their production became quite diminished. In the middle of the Middle Horizon, these systems appeared to have begun to diversify their lithic technology in producing blades for export. The St. Helena systems appear to have undergone a gradual development which resulted in the competition with Trans-Sierran systems. This "competition" appears to have been maintained continuously until European contact. The effects of the competition from the St. Helena system resulted in a digression of the Trans-Sierran systems in the western foothill region. The products of the St. Helena system appear to be more utilitarian from its inception. This focus may have favored the "success" of this system over the Trans-Sierran systems. Finally, it is important to note that following the principle of least effort (Zipf 1949) the St. Helena system always had a definite advantage in terms of accessibility and continuity in supply, considering the heavy snow covering the Trans-Sierran exchange routes from October to May.

EVOLUTION OF SYSTEMS AND THEIR RESPONSES

The last section demonstrated that the three exchange systems definitely underwent systemic changes. Throughout this time, the region as a whole underwent a continuous population growth and increasing sedentism (Heizer 1964). It does appear that with this densification the Sierran foothills became areas of permanent settlement (Moratto 1972), which resulted in increased demand since much of the foothill region lacks flakeable lithic materials. Too, during the Middle Horizon a major change in lithic technology occurred with the appearance of the bow and arrow, replacing the atlatl.

Each of the systems had to respond to these conditions. It does appear that several important changes are registered in production. It is predicted that utilization exchange should precede luxury exchange. The analysis herein indicates the opposite case. The large obsidian bifaces in an Early Horizon context indicate that obsidian was used for the production of luxury items. The Bodie Hills and Casa Diablo were the main contributors of these items, from evidence supplied by Jackson (1974). However, it is not clear why the St. Helena obsidian source did not supply those areas to the same extent as it did later in time. Perhaps the physical properties or even the opaqueness of St. Helena obsidian restricted its use or attractiveness in the production of large bifaces. It is important to note that the input of luxury (obsidian) items from the St. Helena area would have effectively interfered with the eastward flow of shell products from the Bodega Bay area. Whatever the reason for the limited productivity, it remains an interesting research question.

The growth of the St. Helena system and its utilitarian focus does affect the Trans-Sierran systems. As a response during middle Middle Horizon times, these two systems increase their net productive output of bifaces as well as change their lithic technology to blade production, which provides a more generalized item through a greatly simplified production scheme.

It is still uncertain whether or not the decrease in the production rates of the Bodie Hills and Casa Diablo system in Central California is real. What is perplexing is why these systems appear to have receded in time. Perhaps the growth of the St. Helena system was a response to a diminishing supply rather than direct competition. The resolution of this interesting problem remains to be investigated.

Last of all, the impact of the bow and arrow technology on a lithic resource remains to be determined. Obsidians should have been the preferred lithic material for arrowpoint production. Its low density, hardness, brittleness, and particularly its "flakeability" make it a preferable material over alternative lithics. This change in technology itself would have redefined the regional importance of obsidian formerly used for luxury items to its use in utilitarian ones. It is predicted that the introduction of the bow and arrow technology is directly linked with the major changes observed in the three exchange systems under study.

Finally, the theories behind systemic change are extremely complex. It is important to explore and test procedures which will allow us to document systemic changes. The diversity of California provides a complex backdrop against which processes of evolution and change of prehistoric exchange systems can be _further_ investigated.

Chapter 8

SUMMARY AND CONCLUSIONS

Many results reported in the literature suggest that the variety of hydration models proposed may in fact be the result of inadequate control of temporal association, source specification, or hydration (thermal) environment. This study has definitely demonstrated the need for source-specific hydration rates for increased accuracy of the dating technique. In California where many obsidian sources are present in the archaeological record, sources and artifacts must be characterized by both chemical and multivariate analysis. As formulated herein, the application of the long half-life radionuclide INAA technique of chemical characterization provides the greatest precision and an objective means to characterize obsidian artifacts leading to the routinization of laboratory procedure. As a result, it is now possible on a routine basis to chemically characterize obsidian artifacts. However, depending upon the particular research strategy and the facilities available, two other procedures can be used for source identification. The rapid-scan X-ray fluorescence technique (Jack and Carmichael 1969) or the short half-life radionuclide INAA technique, developed herein, are useful.

Sample selection criteria in hydration rate determination have been shown to be very important. In this study, three sampling criteria were used in selecting the data. Samples were drawn from diverse sites and over the widest time span. As a result of this research, a number of new radiocarbon dates are available for California archaeologists. Of particular interest will be an evaluation of the dates from the western foothills of the Sierra Nevadas.

The empirical source-specific obsidian hydration dating rates are presented in table 1-14 for use and evaluation. After extensive testing, it does appear that the original diffusion model (Friedman and Smith 1960) provides the best physical model and general mathematical description so far of the hydration process for the data assembled here. It still remains to be resolved if that diffusion model holds for all cases.

The results of the induced hydration experiments indicate that a large number of observations must be made in order to control for the variability of the induced process. Perhaps this technique can be used to establish the rank-order of source-specific hydration rates. There appears to be a definite thermal discontinuity in the region of $190^{\circ}C$. Beyond that temperature, the hydration band does not appear in most samples. Perhaps beyond $190^{\circ}C$ an annealing effect begins to operate.

Although the activation energies of the induced hydration rates appear to be greater than those shown by Friedman and Long (1976), the comparison of the induced and empirical source-specific hydration rates shows a similar rank-order of the hydration process among the examined sources.

rank-order of the hydration process among the examined sources. Finally, source-specification definitely improves the accuracy of the obsidian hydration dating technique.

An extensive research project was undertaken to isolate the major intrinsic variables which change the rate of hydration among the obsidian sources. For these purposes, many California obsidian source samples were subjected to an extensive series of physical and chemical analyses in order to determine their intrinsic properties. These properties include the following: density, hardness, degree of crystallization, initial and saturated water concentrations, and chemical composition of eight major oxides. Several additional factors were calculated based on these original properties. Stepwise multiple regression analysis was performed using these properties and factors as independent variables with the activation energies of the induced hydration rates as dependent variables. These tentative results suggest the nature of interaction between the hydration process and obsidian.

Among the most important obsidian sources in California are the St. Helena locality in the northern part of the state and Casa Diablo in the central east. The distribution of artifacts from these sources can be traced over distances of hundreds of miles. The earliest established use of obsidian artifacts dates over 10,000 years ago in the Clear Lake area. About 5,000 years ago obsidian began to be utilized to a greater extent. However, its maximum distribution occurs about 500 years ago.

In part 2 the space-utility function of regional exchange in California is examined in detail. California offers an ideal regional variability. The proposed three-dimensional approach overcomes some of the basic limitations of existing methodology for the study of regional exchange. Ten asymmetric and overlapping systems are examined in terms of three variables. The presence and location of alternative lithic materials appear to determine the extent of specific systems. A spatial correspondence was found between the gradients of systems and the location of prehistoric trails. This finding confirms, as one would expect, that the trails served as lines for exchange. However, the lack of correspondence between ethnolinguistic boundaries suggests that further investigation of this problem with more refined data is required.

Regional exchange systems are particularly interesting because they represent large-scale interaction. The isolation of source-specific exchange systems does require both chronological control and characterization of specific exchange items. Once isolated, the morphological properties of each system suggest that the modes of production and organization vary. These differences may be linked to the response of a system to a number of variables. For example, inter-source interaction appears to influence greatly the general morphological properties of each system. Systems having sources without neighboring or local alternative lithic sources within their boundaries tend to be quite symmetrical. However, systems in close proximity are deformed as a consequence of their interaction. Therefore, it follows that systemic interaction is an important factor in determining the growth and formation of a particular system.

The significance of distance as a measure and variable is not clearly understood. It is suggested that its significance may be a function of the mode of acquisition. In a general linear model based on data from the 10 individual systems, distance was found to be the most predictive variable. However, the effect of population or number of potential consumers on diminishing supply within a system definitely warrants further investigation. Although the location of alternative lithic sources appears to be a definite variable, the distance to the next nearest obsidian source appears to be less significant. This result is not understood considering the morphological deformation observed between two interacting systems.

Through multivariate analysis, it is possible to determine the certain properties of each system. The measurement of the catchment and total population verify the results that the modes of production and internal organization vary for each system. California does offer a setting in which to resolve further systemic parameters and processes.

The mechanisms of formation, stability, and change of exchange systems in space and time are interesting problems. It has been proposed that sedentism and population growth both favor the development of regional exchange systems which should be a stepwise process incorporating more people and exchange items. Changes in the systems are difficult to predict; however, a number of types of change are suggested.

A study of the diachronic production rate of three interrelated exchange systems reveals several important aspects of system formation and response. The diachronic patterns of two systems are almost identical in their characteristics. In time, their productivity begins to diminish. This change may be linked to the effects of a third regional exchange system. For example, the growth of the St. Helena system begins in the middle Middle Horizon of Central California and may have caused a regression of the Bodie Hills and Casa Diablo systems. Through lithic production analysis, it does appear that the obsidian was first exchanged as a luxury item in the form of large ceremonial bifaces. Perhaps, in order to meet the completion of St. Helena which supplied utilitarian items to Central California, the two other systems then began to produce generalized blades for utilitarian purposes. While an exhaustive study of the total distribution of obsidiain artifacts would take years to complete and require the excavation of many more sites, this work may serve as a guide for such studies.

Appendix 1

FURTHER NOTES ON OBSIDIAN SOURCES

Appendix 1

FURTHER NOTES ON OBSIDIAN SOURCES

INTRODUCTION

Since 1970, an effort has been made to continue field investigations of the existing obsidian sources and to evaluate any potential information on "new" sources. The following information serves to update the report on the obsidian sources (Ericson, Hagan, and Chesterman 1976) shown in plate 1-1 whose locations are described in table 1-1.

CALIFORNIA OBSIDIAN SOURCES

1. Obsidian Butte, Imperial County — The dates of the extrusion of the Obsidian Butte outcrops are 16,000-14,000 years B.P. (Muffler and White 1969).

2. Emerald Mountain, Kern County — The reported rhyolite dike has not been located upon subsequent investigation.

3. Jawbone Canyon, Kern County — A green, highly fractured perlite has been located in the sediment of Jawbone Canyon. The actual outcrop has not been located.

4. Sugarloaf, Inyo County (Coso) — No new information.

5. Monache Mountain, Tulare County — No new information.

6. Fish Springs, Inyo County — During a recent visit in 1975, evidence for increased "potting activity" was observed throughout the quarry workshops of this area.

7. Inyo Craters, Mono County — No new information.

8. Mono Craters, Mono County — The eruptive sequence of the Mono Crater volcanic chain has been reexamined (Smith 1973).

9. Glass Mountain, Mono Lake, Mono County (Mono Glass Mountain) — In July 1972, additional obsidian samples were collected in the Sawmill Meadows area (NE 1/4, SE 1/4, Sec. 17, T2S, R30E, MDBM). The full extent of the obsidian outcrops remain unsurveyed.

10. Truman Canyon-West Queen Mine, Mono Lake, Mono County (Benton) No new information.

11. Casa Diablo, Mono County — In September 1972, obsidian samples were collected from three areas: (1) SE 1/4, NW 1/4, Sec. 29, T3S, R28E; (2) NE 1/4, SE 1/4, Sec. 35, T3S, R38E; (3) NW 1/4, SE 1/4, Sec.

22, T3S, R28E. This survey was conducted around the periphery of the obsidian extrusion. From chemical characterization analysis of obsidian artifacts (Jackson 1974; Jack 1976) it appears that this obsidian source was very important for the exchange systems of the Late Horizon of Central California.

12. Bodie Hills, Mono County — In July 1974, a survey of the Bodie Hills obsidian quarry, designated as 4-Mno-612, was conducted. A subsequent quarry production analysis of this source suggests that it was extremely important for the exchange systems of the Middle Horizon of Central California (Singer and Ericson 1977).

13. Levitt Peak, Tuolumne County — Three attempts in the field have been made to locate and survey this reported obsidian source. Prior to the late August 1975 snowstorm, which aborted the most recent attempt to explore Levitt Peak, a dome-like structure was observed on the south side of Levitt Peak, using a stereoscope and paired aerial photographs.

14. Deer Creek, Tehama County — A thorough survey of the drainage of Butte Mountain by Tim Hagan did not indicate the presence of an obsidian source on this mountain (Hagan, personal communication). The location of the obsidian source, exploited by the Yahi Indians, is still unidentified (Pope 1918; Heizer and Treganza 1944).

15. Jess Valley, Modoc County — No new information.

16. Cowhead Lake, Surprise Valley, Modoc County — No new information.

17. 8-Mile Creek, Surprise Valley, Modoc County — This source is chemically undifferentiated from Cowhead Lake obsidian source, as shown by characterization analysis.

18. Buck Mountain, Surprise Valley, Modoc County — This obsidian source is being destroyed by modern mining and rockhound groups. Supposedly, the mottled brown obsidian is being distributed nationally and internationally to Japan.

19. Fandango Valley, Surprise Valley, Modoc County — No new information.

20. Sugarhill, Surprise Valley, Modoc County — No new information.

21. Steele Swamp, Modoc County — No new information.

22. Dacite-Rhyolite Composite Flow, Glass Mountain, Medicine Lake, Modoc County (Modoc Glass Mt., Dacite) — No new information.

23. Rhyolite Obsidian Flow, Glass Mountain, Medicine Lake, Siskiyou County (Modoc Glass Mt., Rhyolite) — A sample of this flow has been adopted as the National Obsidian Standard by the U.S. Bureau of Standards (Friedman personal communication). A study of the homogeneity of this composite flow is reported by Wright (1974).

24. Cougar Butte, Medicine Lake, Siskiyou County — No new information.

25. Medicine Lake Glass Flow, Medicine Lake, Siskiyou County — No new information.

26. Little Glass Mountain, Medicine Lake, Siskiyou County — No new information.

27. Grasshopper Flat, Medicine Lake, Siskiyou County — Chemical analysis and other data is described (Parks and Tieh 1966).

28. Winters, Solano County — No new information.

29. Borax Lake, Clear Lake, Lake County — The utilization of this source began at least as early as 10,260 ± 140 years B.P. (Ericson and Berger 1974) in accordance with the early artifacts and faunal evidence of the nearby Borax Lake site presented by Meighan and Haynes (1970). A study for the chemical homogeneity of this source suggests that there is a large variance among analyzed samples (Bowman, Asaro, and Perlman 1973). In addition, the chemical data is statistically overlapped with the adjacent Mt. Konocti obsidian source.

30. Mt. Konocti, Clear Lake, Lake County — This obsidian source was utilized as early as 10,260 ± 140 years B.P. (Ericson and Berger 1974). In fact, it is quite interesting to note the utilization of both the Borax Lake and Mt. Konocti obsidian sources at that time. The aforementioned study by Bowman, Asaro, and Perlman (1973) suggests that there is statistical overlap of the chemical data with the Borax Lake source. A recent geological study of this obsidian outcrop has been made (McNitt 1968).

31. Glass Mountain, St. Helena, Napa County (Napa Glass Mt.) — Several chemical analyses of this obsidian source have appeared in the literature (Jack and Carmichael 1969; Bowman, Asaro, and Perlman 1973; Jackson 1975; Jack 1976; Friedman and Long 1976). Chemical characterization analyses of obsidian artifacts suggest that this obsidian source was quite important during the Late Horizon of Central California (Jackson 1975; Jack 1976).

32. Annadel Farms, Kenwood, Sonoma County (Annadel) — The perlite mining operations have ceased; the obsidian source and surrounding area has become a county park administered by Sonoma County (Kaufman personal communication).

33. Rustler Canyon, San Bernardino County (NE 1/4, NW 1/4, Sec. 16, T11N, R15E, SBBM, Mid Hills quad.) — This newly located source of obsidian, which occurs in the form of pebble-size "Apache tears," is located in Rustler Canyon near the archaeological site, SBr-288 (Davis 1962). Apparently, this material does not seem to have been utilized for chipped stone tool manufacture as indicated by the low percentage of obsidian within the assemblage of SBr-288 (Davis 1962).

34. Dry Creek, Mono County (SE 1/4, Sec. 31, T2S, R28E, NE 1/4, Sec. 6 and NW 1/4, Sec. 5, T2S, R28E, MDBM, Mt. Morrison quad.) — This newly located source of obsidian and 20 square mile obsidian quarry is within the Dry Creek drainage, $3\frac{1}{2}$ miles north of Casa Diablo Hot Springs and 1 mile southeast of Lookout Mt. The area was surveyed in 1974 by T. Balint, B. Miller, and R. Rockwell. The obsidian samples and survey notes were provided by the Inyo National Forest, 2957 Birch Street, Bishop, California 93514.

35. Shoshone, Inyo County — Gray porphyritic-obsidian-perlite outcrops extensively in the Shoshone area. Four areas were surveyed: Dublin Hills (SW 1/4, NW 1/4, Sec. 23, T22N, RGE, MDBM, Shoshone quad.); Salisbury Pass (SE 1/4, SW 1/4, Sec. 16, T21N, R5E, MDBM, Shoshone quad.); Jubilee Pass (NW 1/4, NE 1/4, Sec. 19, T21N, R4E, MDBM, Confidence Hills quad.); Charley Brown (Sec. 4, T21N, R7E, MDBM, Tecopa). It is not expected nor observed that this material was used prehistorically for tool manufacture, due to the porphyritic texture and perlitic structure of the obsidian.

36. Kelly Mountain, Lassen County (T29N, R6E, MDBM) — Kelly Mountain (Jackson 1975:48) may be the elusive "Source X" (Jack 1976). The "Source X," designated by UCLA, refers to a suite of samples which were collected in Lassen County by an old prospector. The samples exchanged hands at least three times before arriving at UCLA, courtesy of T. A. Hagan. Unfortunately, the provenience of the samples remains unknown. Whether Kelly Mountain (Jackson 1974:48), "Source X," UCB (Jack 1976), and "Source X,", UCLA, are one and the same source remains to be resolved, hopefully by future fieldwork and chemical characterization analysis.

CALIFORNIA SOURCES (NATURAL GLASSES)

G1. El Toro Glass, Orange County — Investigation of the reported location near Leisure Hills Stables revealed a "glassy" chert reported by Nesselrod (personal communication).

G2. Grimes Canyon Fused Shale, Ventura County — No new information.

G3. Cuyama Glass, Santa Barbara County — There is no natural glass suitable for artifact manufacture associated with the "triple" basalt flows (Clifton 1967) or other extrusive rocks in the Caliente Range (Vedder personal communication). On the other hand, siliceous shale and chert of Monterey Shale and shaly units in the Santa Margarita Formation in the Cuyama Valley were commonly used for tools and points (Vedder personal communication). Most likely these units were mistakenly reported as natural glass (Ericson, Hagan, and Chesterman 1976).

G4. Shell Beach Zeolitized Tuffs, San Luis Obispo County — No new information.

WESTERN NEVADA SOURCES

N1. Mount Hicks, Mineral County — Three successive surveys in 1972, 1974, and 1975 were made of the Mt. Hicks obsidian source. In 1972, several samples were removed from a local chipping station and potential hunting camp (Sec. 21, T5N, R29E) as samples representative of this source. In 1974, geological samples were collected on the southeast side of Mt. Hicks (NE 1/4, NE 1/4, Sec. 19, T5N, R29E), where extensive debitage was observed. In 1975, geological samples were collected from the summit of Mt. Hicks (NW 1/4 SW 1/4, Sec. 24,

T5N, R28E). From these initial surveys it appears that the major concentration of debitage within quarry-workshop areas may exceed five square miles.

N2. Duck Flat, Washoe County — No new information.

N3. Long Valley, Washoe County — A survey of this dispersed obsidian source was made (Ragir and Lancaster 1966).

N4. Fletcher, Mineral County — (NE 1/4, NW 1/4, NW 1/4, Sec. 24, T6N, R27E, Pine Grove Hills quad.) — This newly located obsidian source is a small outcrop of less than an acre of obsidian nodules and some debitage. Its location along the hypothenuse, formed by Mt. Hicks and Pine Grove Hills, of northeasterly-oriented right triangle formed by Bodie Hills, is approximately equidistant from each of these three sources.

N5. Pine Grove Hills, Mineral County (NE 1/4, NW 1/4, Sec. 31, T9N, R26E, Pine Grove Hills quad.) — This newly located obsidian source appears to be the least disturbed obsidian quarry; no modern material remains were observed. The quarry workshop of this small obsidian source is located within 200 feet of the summit. It appears as if the supply of obsidian nodules was totally exhausted prehistorically. Little, if any, obsidian or debitage has been transported downslope within the drainage system, which suggests that the formation of this source may have been quite recent.

SOUTHERN OREGON SOURCES

O1. Beatty's Butte, Lake County — No new information.

O2. Glass Mountain, Lake County — No new information.

O3. Glass Butte, Lake County — No new information.

O4. The Waibel Sources — As a result of the correspondence between the author and Al Waibel, then associated with the Earth Sciences Department, Portland State University, Oregon, in 1973, the following list of obsidian sources in the Southern Oregon area has been compiled (Waibel personal communication):

    Skull Creek (Sec. 2, T22S, R29E, WBM)
    Catnip Mountain (Sec. 7, T46N, R23E, WBM)
    Drews Creek (Sec. 18, T38S, R16E, WBM)
    Burns Butte (Sec. 20, T23S, R30E, WBM)
    Dooly Mountain (Sec. 32, T11S, R40E, WBM)
    Horse Mountain (Sec. 8, T28S, R22E, WBM)
    Quartz Mountain (Sec. 26, T22S, R15E, WBM)
    Little Bear Creek (Sec. 21, T16S, R33E, WBM)
    Cougar Mountain (Sec. 14, T25S, R15E, WBM)
    Reilly (Sec. 18, T24S, R27E, WBM)
    East Butte (Sec. 13, T22S, R14E, WBM)
    Clarno (no data, WBM)
    Liberty, Idaho (T8N, R2E, BBM)

Appendix 2

PRIMARY DATA SET FOR SOURCE-SPECIFIC
OBSIDIAN HYDRATION RATES

Appendix 2

PRIMARY DATA SET FOR SOURCE-SPECIFIC
OBSIDIAN HYDRATION RATES

This is a list of column headings and other abbreviations used in appendix 2.

No. = Sample number.

Source = Chemical assigned source.

Tech. = Chemical analysis technique.

Anal. No. = UCLA INAA Logbook number.

Prog. No. = Computer program.

CC2 = Central California, version 2.
CC4 = Central California, version 4.
LC1 = Long half-life California, version 1.

Prob. = Probability of correct assignment, which varies from 0 to 1.

D^2 = Mahalabinos " d-squared" distance.

Site = Archaeological site number, county abbreviated.

Unit = Intrasite excavation unit.

Level = Excavation level.

Coll. = Institution housing the collection.

Art. No. = Institutional number of artifact.

Sub. = Last name of submitter of the sample.

Lab. No. = Radiocarbon Laboratory and sample number.

C^{14} Date = Radiocarbon date and 2σ deviation in years before present.

Material = Sample material.

Cal. Date = Bristlecone pine corrected date A.D./B.C.

Reference = Reference to radiocarbon date.

Assoc. = Type of association between the obsidian samples and radiocarbon date (cf. criteria specified in chapter 1).

Lab No. = UCLA Obsidian Hydration Laboratory number.

Hydra. = Hydration measurement in microns.

Mean = Group mean of the hydration measurements of two or more obsidian artifacts, if available.

Analyst = Observer of the hydration measurement.

Appendix 2A

CHEMICAL CHARACTERIZATION AND SAMPLE
PROVENIENCE DATA

Appendix 2A

CHEMICAL CHARACTERIZATION AND SAMPLE PROVENIENCE DATA

No.	Source	Tech.	Anal. No.	Prog. No.	Prob.	D^2	Site	Unit	Level	Coll.	Art. No.	Sub.
1	Annadel	Short	894	CC4	1.0	4.8	CCo-30	Q-40	18-24"	SDA	2	Frederickson
2	Annadel	Short	906	CC4	1.0	2.9	CCo-30	Q-40	36-42"	SDA	2	Frederickson
3	Annadel	Short	1194	CC4	1.0	2.0	Mrn-152	Bur. 5	90cm	---	---	Clewlow
4	Annadel	Short	1197	CC4	1.0	13.8	Mrn-152	Bur. 5	90cm	---	---	Clewlow
5	Annadel	Short	1202	CC4	1.0	0.2	Mrn-152	Bur. 4	85-90cm	---	---	Clewlow
6	Annadel	Short	1205	CC4	1.0	0.2	Mrn-152	Bur. 4	85-90cm	---	---	Clewlow
7	Annadel	Short	756	CC2	11.0	0.2	Ala-307	H-4	202"	UCB	1-123421	JEE
8	Annadel	Short	678	CC2	0.95	24.3	Mrp-105	Pit 22C	18-24"	UCLA	335-1712	JEE
9	Annadel	Short	679	CC2	0.80	30.0	Mrp-105	Pit 22C	18-24"	UCLA	335-1714	JEE
10	Annadel	Short	681	CC2	0.99	23.0	Mrp-105	Pit 22C	18-24"	UCLA	335-1716	JEE
11	Annadel	Short	686	CC2	0.99	27.3	Mrp-105	Pit 22C	18-24"	UCLA	335-1722	JEE
12	Bodie Hills	Short	693	CC2	0.971	13.7	Mrp-105	Pit 22C	36-42"	UCLA	335-1758	JEE
13	Bodie Hills	Short	761	CC2	0.882	5.8	Sjo-142	Bur. 18	28"	UCB	1-48802	Bennyhoff
14	Bodie Hills	Short	770	CC2	0.851	3.2	Sjo-68	Crem. 1	47"	UCB	1-73243E	Bennyhoff
15	Bodie Hills	Short	811	CC2			Son-518	H.F.	Feature 1	SDA	73-9-004	Upson
16	Bodie Hills	Short	812	CC4	0.876	18.0	Son-518	H.F.	Feature 1	SDA	73-9-629	Upson
17	Bodie Hills	Short	813	CC4			Son-518	H.F.	Feature 1	SDA	73-9-631	Upson
18	Bodie Hills	Long	1659	LCI	0.818	9.0	Ala-307	F-5	150"	UCB	1-122880	JEE
19	Bodie Hills	Long	1663	LCI	0.668	11.0	Cal-99	Pit 3	14"	UCB	1-139113	JEE
20	Bodie Hills	Long	1667	LCI	0.538	12.0	Cal-99	Pit 3	30"	UCB	1l139114	JEE
21	Bodie Hills	Long	1669	LCI	0.651	13.0	Cal-99	Pit 3	29"	UCB	1-139116	JEE
22	Bodie Hills	Long	1671	LCI	0.914	18.0	Cal-99	Pit 3	24"	UCB	1-139119	JEE
23	Bodie Hills	Long	1673	LCI	0.816	13.0	Cal-99	Pit 3	24"	UCB	1-139162	JEE
24	Borax Lake	Short	814	CC4	1.0	35.6	Son-518	H.F.	Feature 1	SDA	74-9-628	Upson
25	Borax Lake	Short	816	CC4	0.998	17.0	Son-518	H.F.	Feature 1	SDA	73-9-006	Upson

(contd.)

Appendix 2A (contd.)

No.	Source	Tech.	Anal. No.	Prog. No.	Prob.	D^2	Site	Unit	Level	Coll.	Art. No.	Art. No.	Sub.
26	Borax Lake	Short	817	CC4	1.0	21.5	Son-518	H.F.	Feature 1	SDA	None		Upson
27	Borax Lake	Short	910	CC4	1.0	18.9	CCo-30	Q-40	36-42"	SDA	6		Frederickson
28	Borax Lake	XRF	992	---	---	---	Lak-380	Bur. 4	---	SDA	M-1		R. King
29	Borax Lake	XRF	993	---	---	---	Lak-380	Bur. 4	---	SDA	M-2		R. King
30	Borax Lake	XRF	994	---	---	---	Lak-380	Bur. 4	---	SDA	M-6		R. King
31	Borax Lake	XRF	997	---	---	---	Lak-380	Bur. 4	---	SDA	m-6		R. King
32	Borax Lake	XRF	---	---	---	---	Lak-380	Bur. 9	---	SDA	5		R. King
33	Borax Lake	Long	1655	LCI	0.998	31	Nev-15	16SW20	12-24"	UCB	1-173134-1		JEE
34	Borax Lake	Long	1658	LCI	0.708	23	Nev-15	22-SE-20	12-24"	UCB	1-173199		JEE
35	Borax Lake	Short	771	CC2	0.934	13.9	Sjo-68	Crem. 1	47"	UCB	1-173243f		JEE
36	Borax Lake	XRF	---	---	---	---	Lak-261	0	90-96"	SDA	11-2		Findlow
37	Borax Lake	XRF	---	---	---	---	Lak-261	0	90-96"	SDA	11-8		Findlow
38	Borax Lake	XRF	---	---	---	---	Lak-261	0	90-96"	SDA	11-5		Findlow
39	Borax Lake	XRF	---	---	---	---	Lak-261	0	90-96"	SDA	11-3		Findlow
40	Borax Lake	XRF	---	---	---	---	Lak-261	0	90-96"	SDA	11-6		Findlow
41	Borax Lake	XRF	---	---	---	---	Lak-261	0	90-96"	SDA	11-13		Findlow
42	Borax Lake	XRF	---	---	---	---	Lak-261	0	90-96"	SDA	11-11		Findlow
43	Borax Lake	XRF	---	---	---	---	Lak-261	0	90-96"	SDA	11-1		Findlow
44	Borax Lake	XRF	---	---	---	---	Lak-261	0	90-96"	SDA	11-4		Findlow
45	Borax Lake	XRF	---	---	---	---	Lak-261	0	90-96"	SDA	11-9		Findlow
46	Borax Lake	XRF	---	---	---	---	Lak-261	0	90-96"	SDA	11-7		Findlow
47	Borax Lake	XRF	---	---	---	---	Lak-261	0	48-60"	SDA	261-S-48-4		Findlow
48	Borax Lake	XRF	---	---	---	---	Lak-261	0	48-60"	SDA	261-S-48-1		Findlow
49	Borax Lake	XRF	---	---	---	---	Lak-261	0	48-60"	SDA	261-S-48-2		Findlow
50	Borax Lake	XRF	---	---	---	---	Lak-261	0	48-60"	SDA	261-S-48-3		Findlow
51	Borax Lake	XRF	---	---	---	---	Lak-261	0	48-60"	SDA	261-S-54-1		Findlow

(contd.)

Appendix 2A (contd.)

No.	Source	Tech.	Anal. No.	Prog. No.	Prob.	D^2	Site	Unit	Level	Coll.	Art. No.	Sub.
52	Borax Lake	XRF	---	---	---	---	Lak-261	0	48-60"	SDA	261-S-54-2	Findlow
53	Borax Lake	XRF	---	---	---	---	Lak-261	0	48-60"	SDA	261-S-54-3	Findlow
54	Borax Lake	XRF	---	---	---	---	Lak-261	0	48-60"	SDA	261-S-54-4	Findlow
55	Borax Lake	XRF	---	---	---	---	Lak-261	0	48-60"	SDA	261-S-54-5	Findlow
56	Buck Mt.	Long	54c	---	---	---	4SK4	N100E86	3	UO	54c	Sampson
57	Buck Mt.	Long	54d	---	---	---	4SK4	N100E86		UO	54d	Sampson
58	Buck Mt.	Long	58b	---	---	---	4SK4	N100E86	5	UO	58b	Sampson
59	Buck Mt.	Long	58d	---	---	---	4SK4	N100E86		UO	58d	Sampson
60	Buck Mt.	Long	59b	---	---	---	4SK4	N100E86	6	UO	59b	Sampson
61	Buck Mt.	Long	62a	---	---	---	4SK4	N100E86	6	UO	62a	Sampson
62	Buck Mt.	Long	62j	---	---	---	4SK4	N100E86		UO	62j	Sampson
63	Buck Mt.	Long	64h	---	---	---	4SK4	N100E86	7a	UO	64h	Sampson
64	Buck Mt.	Long	65f	---	---	---	4SK4	N100E86	7b	UO	65f	Sampson
65	Buck Mt.	Long	71f	---	---	---	4SK4	N100E86	8	UO	71f	Sampson
66	Buck Mt.	Long	72i	---	---	---	4SK4	N100E86	9a	UO	72i	Sampson
67	Buck Mt.	Long	73d	---	---	---	4SK4	N100E86	9a	UO	73d	Sampson
68	Buck Mt.	Long	73e	---	---	---	4SK4	N100E86	9a	UO	73e	Sampson
69	Buck Mt.	Long	74g	---	---	---	4SK4	N100E86	9b	UO	74g	Sampson
70	Buck Mt.	Long	74h	---	---	---	4SK4	N100E86	9b	UO	74h	Sampson
71	Buck Mt.	Long	75h	---	---	---	4SK4	N100E86	9b	UO	75h	Sampson
72	Buck Mt.	Long	76d	---	---	---	4SK4	N100E86	10	UO	76d	Sampson
73	Buck Mt.	Long	76f	---	---	---	4SK4	N100E86	10	UO	76f	Sampson
74	Buck Mt.	Long	89c	---	---	---	4SK4	N84E44	2	UO	89c	Sampson
75	Buck Mt.	Long	90b	---	---	---	4SK4	N84E44	2	UO	90b	Sampson
76	Buck Mt.	Long	90c	---	---	---	4SK4	N84E44	2	UO	90c	Sampson
77	Casa Diablo	Assumed	---	---	---	---	Mno-382	Pit 29	6-12"	UCLA	436-2585	JEE
78	Casa Diablo	Assumed	---	---	---	---	Mno-382	Pit 16	6-12"	UCLA	436-1955	JEE
79	Casa Diablo	Assumed	---	---	---	---	Mno-382	Pit 16	6-12"	UCLA	436-1968	JEE

(contd.)

Appendix 2A (contd.)

No.	Source	Tech.	Anal. No.	Prog. No.	Prob.	D^2	Site	Unit	Level	Coll.	Art. No.	Sub.
80.	Casa Diablo	Assumed	---	---	---	---	Mno-382	Pit 17	6–12"	UCLA	436-1945	JEE
81	Casa Diablo	Assumed	---	---	---	---	Mno-382	Pit 14	6–12"	UCLA	436-1279	JEE
82	Casa Diablo	Assumed	---	---	---	---	Mno-382	Pit 26	6–12"	UCLA	436-7170	JEE
83	Casa Diablo	Assumed	---	---	---	---	Mno-382	Pit 26	6–12"	UCLA	436-6919	JEE
84	Casa Diablo	Short	695	CC2	0.939	5.7	Mrp-105	Pit 22C	48–54"	UCLA	436-1766	JEE
85	Casa Diablo	Short	691	CC2	0.977	5.0	Mrp-105	Pit 22C	36–42"	UCLA	335-1754	JEE
86	Casa Diablo	Short	648	S2	0.996	23.3	Riv-463	Area 2	60–70 cm	UCLA	G4	Wilke
87	Casa Diablo	Short	1090	S4	0.992	2.2	Lan-264	Bur. 35		UCLA	Bur. 35	Meighan
88	Casa Diablo	Short	1091	S4	0.992	3.0	Lan-264	A19	84–90"	UCLA	573-1354 A	Meighan
89	Casa Diablo	Short	1562	CC6	0.604	2.8	Mad-179	H.P. 15	120–140cm	---	1	Stickel
90	Casa Diablo	Short	1564	CC6	0.583	8.4	Mad-179	H.P. 15	120–140cm	---	3	Stickel
91	Casa Diablo	Short	1567	CC6	0.669	1.6	Mad-179	H.P. 15	120–140cm	---	6	Stickel
92	Casa Diablo	Short	1568	CC6	0.925	3.9	Mad-179	H.P. 15	120–140cm	---	7	Stickel
93	Casa Diablo	Short	1569	CC6	0.952	1.1	Mad-179	H.P. 15	120–140cm	---	8	Stickel
94	Casa Diablo	Short	1570	CC6	0.999	2.1	Mad-179	H.P. 15	120–140cm	---	9	Stickel
95	Casa Diablo	Short	1571	CC6	0.571	3.5	Mad-179	H.P. 15	120–140cm	---	10	Stickel
96	Casa Diablo	Short	1572	CC6	0.921	4.1	Mad-179	H.P. 15	120–140cm	---	11	Stickel
97	Casa Diablo	Short	1574	CC6	0.877	1.9	Mad-179	H.P. 15	120–140cm	---	13	Stickel
98	Casa Diablo	Short	1577	CC6	0.903	6.6	Mad-179	H.P. 15	140–160cm	---	2	Stickel
99	Casa Diablo	Short	1583	CC6	0.943	1.7	Mad-179	H.P. 15	140–160cm	---	8	Stickel
100	Casa Diablo	Short	1584	CC6	0.884	1.7	Mad-179	H.P. 15	140–160cm	---	9	Stickel
101	Casa Diablo	Short	1585	CC6	0.954	3.2	Mad-179	H.P. 15	140–160cm	---	10	Stickel
102	Casa Diablo	Short	1586	CC6	0.941	0.8	Mad-179	H.P. 15	140–160cm	---	11	Stickel
103	Casa Diablo	Short	1565	CC6	0.786	0.9	Mad-179	H.P. 15	120–140cm	---	4	Stickel
104	Casa Diablo (Sec. 29)	Long	1657	LC1	1.0	420	Nev-15	16 SW20	12–24"	UCB	1-173134-3	JEE
105	Casa Diablo (Sec. 29)	Long	1668	LC1	0.510	9	Cal-99	Pit 3	24"	UCB	1-139115	JEE

(contd.)

Appendix 2A (contd.)

No.	Source	Tech.	Anal. No.	Prog. No.	Prob.	D^2	Site	Unit	Level	Coll.	Art. No.	Sub.
106	Casa Diablo (Sec. 29)	Long	1672	LC1	0.764	8	Cal-99	Pit 3	22''	UCB	1-139120	JEE
107	Casa Diablo (Sec. 29)	Long	1705	LC1	0.619	6	Mno-	Unit 4	20-40cm	---	---	N. Leonard
108	Casa Diablo (Sec. 35)	Long	1706	LC1	0.777	9	Mno-	Unit 4	20-40cm	---	---	N. Leonard
109	Casa Diablo (Sec. 35)	Long	1707	LC1	0.433	9	Mno-	Unit 4	20-40cm	---	---	N. Leonard
110	Casa Diablo (Sec. 35)	Long	1708	LC1	0.642	9	Mno-	Unit 4	20-40cm	---	---	N. Leonard
111	Casa Diablo (Sec. 35)	Long	1709	LC1	0.489	8	Mno-	Unit 4	20-40cm	---	---	N. Leonard
112	Casa Diablo (Sec. 35)	Long	1710	LC1	0.492	8	Mno-	Unit 4	20-40cm	---	---	N. Leonard
113	Casa Diablo (Sec. 22)	Long	1714	LC1	0.544	10	Mno-	Unit 4	20-40cm	---	---	N. Leonard
114	Casa Diablo (Sec. 35)	Long	1715	LC1	0.596	8	Mno-	Unit 4	20-40cm	---	---	N. Leonard
115	Casa Diablo (Sec. 22)	Long	1716	LC1	0.565	10	Mno-	Unit 4	20-40cm	---	---	N. Leonard
116	Casa Diablo (Sec. 22)	Long	1717	LC1	0.495	10	Mno-	Unit 4	20-40cm	---	---	N. Leonard
117	Casa Diablo (Sec. 29)	Long	1718	LC1	0.840	7	Mno-	Unit 4	20-40cm	---	---	N. Leonard
118	Casa Diablo (Sec. 35)	Long	1719	LC1	0.512	10	Mno-	Unit 4	20-40cm	---	---	N. Leonard
119	Casa Diablo (Sec. 29)	Long	1720	LC1	0.843	6	Mno-	Unit 4	20-40cm	---	---	N. Leonard

(contd.)

Appendix 2A (contd.)

No.	Source	Tech.	Anal. No.	Prog. No.	Prob.	D^2	Site	Unit	Level	Coll.	Art. No.	Sub.
120	Casa Diablo (Sec. 35)	Long	1721	LC1	0.811	11	Mno-	Unit 4	20-40cm	---	---	N. Leonard
121	Casa Diablo (Sec. 35)	Long	1722	LC1	0.788	10	Mno-	Unit 4	20-40cm	---	---	N. Leonard
122	Casa Diablo (Sec. 35)	Long	1723	LC1	0.506	7	Mno-	Unit 4	20-40cm	---	---	N. Leonard
123	Casa Diablo (Sec. 29)	Long	1724	LC1	0.772	8	Mno-	Unit 4	20-40cm	---	---	N. Leonard
124	Casa Diablo (Sec. 29)	Long	1725	LC1	0.795	9	Mno-	Unit 4	20-40cm	---	---	N. Leonard
125	Casa Diablo (Sec. 35)	Long	1726	LC1	0.556	11	Mno-	Unit 4	20-40cm	---	---	N. Leonard
126	Casa Diablo (Sec. 35)	Long	1727	LC1	0.845	9	Mno-	Unit 4	20-40cm	---	---	N. Leonard
127	Casa Diablo (Sec. 35)	Long	1728	LC1	0.856	9	Mno-	Unit 4	20-40cm	---	---	N. Leonard
128	Casa Diablo (Sec. 35)	Long	1729	LC1	0.829	8	Mno-	Unit 4	20-40cm	---	---	N. Leonard
129	Casa Diablo (Sec. 35)	Long	1730	LC1	0.891	24	Mno-	Unit 4	20-40cm	---	---	N. Leonard
130	Coso	Short	575	S2	0.978	2.5	LAn-324	C-6	0-15 cm	---	W66-1225	N. Leonard
131	Coso	Short	579	S2	0.998	2.4	LAn-324	E-10	45-60cm	---	W66-153	Singer
132	Coso	Short	581	S2	1.0	3.2	LAn-324	EF-12	0-15cm	---	W66-447	Singer
133	Coso	Short	582	S2	1.0	0.7	LAn-324	E-10	0-15cm	---	W66-509	Singer
134	Coso	Short	584	S2	0.977	6.0	LAn-324	Knoll-S7/W1	0-15cm	---	W66-1209	Singer
135	Coso	Short	586	S2	0.987	3.5	LAn-324	Knoll S1/E2	30-45cm	---	W66-1231	Singer
136	Coso	Short	588	S2	0.727	10.4	LAn-324	Area 2	0-30cm	---	W66-1589	Singer
137	Coso	Short	627	S2	0.995	35.3	Riv-463	A24	30-40cm	---	D-2	Wilke

(contd.)

Appendix 2 A (contd.)

No.	Source	Tech.	Anal. No.	Prog. No.	Prob.	D^2	Site	Unit	Level	Coll.	Art. No.	Sub.
138	Coso	Short	1101	S4	1.0	0.3	LAn-264	A23	74–78"	UCLA	573-1445	Meighan
139	Coso	Short	1141	S4	1.0	1.7	LAn-264	A23	80–86"	UCLA	573-2021	Meighan
140	Coso	Short	1142	S4	1.0	0.9	LAn-264	A23	80–86"	UCLA	573-2022	Meighan
141	Coso	Short	1143	S4	1.0	2.9	LAn-264	Pit 3C70	82–86"	---	573-2014	Meighan
142	Coso	Short	1587	S3	---	---	Barrows	Pit 3C70	1.5m	---	4C70-1	E. L. Davis
143	Coso	Short	1588	S3	---	---	Barrows	Pit 3C70	1.5m	---	4C70-2	E. L. Davis
144	Coso	Short	1589	S3	---	---	Barrows	Pit 3C70	1.5m	---	4C70-3	E. L. Davis
145	Coso	Short	1590	S3	---	---	Barrows	Pit 3C70	1.5 m	---	4C70-4	E. L. Davis
146	Coso	Short	1591	S3	---	---	Barrows	Pit 3C70	1.5 m	---	4C70-5	E. L. Davis
147	Coso	Short	1602	S3	---	---	Barrows	Pit 3C70	1.5m	---	4C70	E. L. Davis
148	Coso	Long	1611	LC1	1.0	380	Iny-372	Rose Sp.	60–64"	UCB	1-188101	JEE
149	Coso	Long	1612	LC1	1.0	618	Iny-372	Rose Sp.	60–64"	UCB	1-188102	JEE
150	Coso	Long	1613	LC1	1.0	786	Iny-372	Rose Sp.	60–64"	UCB	1-188103	JEE
151	Coso	Long	1614	LC1	1.0	15	Iny-372	Rose Sp.	60–64"	UCB	1-188104	JEE
152	Coso	Long	1615	LC1	1.9	319	Iny-372	Rose Sp.	60–64"	UCB	1-188105	JEE
153	Coso	Long	1616	LC1	1.0	21	Iny-372	Rose Sp.	60–64"	UCB	1-188106	JEE
154	Coso	Long	1617	LC1	1.0	61	Iny-372	Rose Sp.	60–64"	UCB	1-188107	JEE
155	Coso	Long	1618	LC1	1.0	83	Iny-372	Rose Sp.	60–64"	UCB	1-188108	JEE
156	Coso	Long	1621	LC1	1.0	747	Iny-372	Rose Sp.	60–64"	UCB	1-188161	JEE
157	Coso	Long	1623	LC1	1.0	471	Iny-372	Rose Sp.	60–64"	UCB	1-188166	JEE
158	Coso	Long	1624	LC1	1.0	708	Iny-372	Rose Sp.	60–64"	UCB	1-188167	JEE
159	Coso	Long	1625	LC1	1.0	122	Iny-372	Rose Sp.	60–64"	UCB	1-188168	JEE
160	Coso	Long	1626	LC1	1.0	92	Iny-372	Rose Sp.	60–64"	UCB	1-188169	JEE
161	Coso	Long	1627	LC1	1.0	950	Iny-372	Rose Sp.	60–64"	UCB	1-188170	JEE
162	Coso	Long	1628	LC1	1.0	350	Iny-372	Rose Sp.	60–64"	UCB	1-188171	JEE
163	Coso	Long	1629	LC1	1.0	281	Iny-372	Rose Sp.	60–64"	UCB	1-188172	JEE
164	Coso	Long	1632	LC1	1.0	52	Iny-372	Rose Sp.	60–64"	UCB	1-188176	JEE

(contd.)

Appendix 2A (contd.)

No.	Source	Tech.	Anal. No.	Prog. No.	Prob.	D²	Site	Unit	Level	Coll.	Art. No.	Sub.
165	Coso	Long	1633	LC1	1.0	39	Iny-372	D-3	84–90"	UCB	1-144791A	JEE
166	Coso	Long	1634	LC1	1.0	50	Iny-372	D-3	84–90"	UCB	1-144791B	JEE
167	Coso	Long	1635	LC1	1.0	910	Iny-372	D-3	96–102"	UCB	1-144785A	JEE
168	Coso	Long	1636	LC1	1.0	362	Iny-372	D-3	96–102"	UCB	1-144785B	JEE
169	Coso	Long	1637	LC1	1.0	391	Iny-372	D-3	96–102"	UCB	1-144785C	JEE
170	Coso	Long	1638	LC1	1.0	45	Iny-372	D-3	96–102"	UCB	1-144785D	JEE
171	Coso	Long	1639	LC1	1.0	401	Iny-372	D-3	96–102"	UCB	1-144785E	JEE
172	Coso	Long	1640	LC1	1.0	366	Iny-372	D-3	96–102"	UCB	1-144785E	JEE
173	Coso	Long	1641	LC1	1.0	454	Iny-372	D-3	96–102"	UCB	1-144785G	JEE
174	Coso	Long	1642	LC1	1.0	417	Iny-372	D-3	96–102"	UCB	1-144785H	JEE
175	Coso	Long	1643	LC1	1.0	429	Iny-372	D-3	96–102"	UCB	1-144785I	JEE
176	Coso	Long	1644	LC1	1.0	411	Iny-372	D-3	96–102"	UCB	1-144785J	JEE
177	Coso	Long	1645	LC1	1.0	407	Iny-372	D-3	102–108"	UCB	1-144790A	JEE
178	Coso	Long	1646	LC1	1.0	371	Iny-372	D-3	102–108"	UCB	1-144790B	JEE
179	Coso	Long	1647	LC1	1.0	39	Iny-372	D-3	102–108"	UCB	1-144790C	JEE
180	Coso	Long	1648	LC1	1.0	37	Iny-372	D-3	102–108"	UCB	1-144790D	JEE
181	Coso	Long	1649	LC1	1.0	386	Iny-372	D-3	102–108"	UCB	1-144790E	JEE
182	Coso	Long	1650	LC1	1.0	371	Iny-372	D-1	90–102"	UCB	1-144793A	JEE
183	Coso	Long	1651	LC1	1.0	341	Iny-372	D-1	90–102"	UCB	1-144793B	JEE
184	Coso	Long	1731	LC1	1.0	38	Iny-222	C-8	11–12"	UCB	1-130480-1	JEE
185	Coso	Long	1732	LC1	1.0	45	Iny-222	C-8	11–12"	UCB	1-130480-2	JEE
186	Coso	Long	1733	LC1	1.0	30	Iny-222	C-8	11–12"	UCB	1-130480-3	JEE
187	Coso	Long	1735	LC1	1.0	38	Iny-222	C-8	11–12"	UCB	1-130480-5	JEE
188	Coso	Long	1736	LC1	1.0	52	Iny-222	C-8	11–12"	UCB	1-130480-6	JEE
189	Coso	Long	1737	LC1	1.0	293	Iny-222	C-8	11–12"	UCB	1-130480-7	JEE
190	Coso	Long	1738	LC1	1.0	306	Iny-222	C-8	11–12"	UCB	1-130480-8	JEE
191	Coso	Long	1741	LC1	1.0	296	Iny-222	C-8	11–12"	UCB	1-130480-10	JEE (contd.)

Appendix A2 (contd.)

No.	Source	Tech.	Anal. No.	Prog. No.	Prob.	D^2	Site	Unit	Level	Coll.	Art. No.	Sub.
192	Coso	Long	1746	LC1	1.0	292	Iny-222	5-LI	0-6"	UCB	1-202489	JEE
193	Coso	Long	1747	LC1	1.0	309	Iny-222	5-LI	0-6"	UCB	1-202489-1	JEE
194	Coso	Long	1748	LC1	1.0	331	Iny-222	5-LI	0-6"	UCB	1-202689-2	JEE
195	Coso	Long	1749	LC1	1.0	307	Iny-222	5-LI	0-6'	UCB	1-202689-3	JEE
196	Coso	Long	1750	LC1	1.0	330	Iny-222	5-LI	0 6"	UCB	1-202689-4	JEE
197	Coso	Short	787	S2			SBCM-128	N4/E4SQC	30-40cm	SBCM	128-008A	White
198	Coso	Short	789	S2			SBCM-128	N4/E4SQC	30-40cm	SBCM	128-008A	White
199	Coso	Short	790	S2			SBCM-128	N4/E4SQD	30-40cm	SBCM	128-009A	White
200	Coso	Short	791	S2			SBCM-128	N4/E4SQD	30-40cm	SBCM	128-009B	White
201	Coso	Short	792	S2			SBCM-128	N4/E4SQD	30-40cm	SBCM	128-009C	White
202	Coso	Short	793	S2			SBCM-128	N6/NN8SQA	30-40cm	SBCM	128-056	White
203	Coso	Short	794	S2			SBCM-128	N6/W8SQC	30-40cm	SBCM	128-0014	White
204	Coso	Short	798	S2			Ora-232	N-14	16-22"	BM	#92	JEE
205	Modoc Glass Mt.	Long	54a	---	---	---	4SK4	N100E86	3	UO	54a	Sampson
206	Modoc Glass Mt.	Long	54f	---	---	---	4SK4	N100E86	3	UO	54f	Sampson
207	Modoc Glass Mt.	Long	54i	---	---	---	4SK4	N100E86	3	UO	54i	Sampson
208	Modoc Glass Mt.	Long	57f	---	---	---	4SK4	N100E86	5	UO	54f	Sampson
209	Modoc Glass Mt.	Long	58i	---	---	---	4SK4	N100E86	5	UO	58i	Sampson
210	Modoc Glass Mt.	Long	59g	---	---	---	4SK4	N100E86	6	UO	59g	Sampson
211	Modoc Glass Mt.	Long	59i	---	---	---	4SK4	N100E86	6	UO	59i	Sampson
212	Modoc Glass Mt.	Long	59k	---	---	---	4SK4	N100E86	6	UO	59k	Sampson
213	Modoc Glass Mt.	Long	60e	---	---	---	4SK4	N100E86	6	UO	60e	Sampson
214	Modoc Glass Mt.	Long	60g	---	---	---	4SK4	N100E86	6	UO	60g	Sampson
215	Modoc Glass Mt.	Long	60h	---	---	---	4SK4	N100E86	6	UO	60h	Sampson
216	Modoc Glass Mt.	Long	60k	---	---	---	4SK4	N100E86	6	UO	60k	Sampson
217	Modoc Glass Mt.	Long	62d	---	---	---	4SK4	N100E86	6	UO	62d	Sampson
218	Modoc Glass Mt.	Long	62h	---	---	---	4SK4	N100E86	6	UO	62h	Sampson

Appendix 2A (contd.)

No.	Source	Tech.	Anal. No.	Prog. No.	Prob.	D^2	Site	Unit	Level	Coll.	Art. No.	Sub.
219	Modoc Glass Mt.	Long	63b	---	---	---	4SK4	N100E86	7a	UO	63b	Sampson
220	Modoc Glass Mt.	Long	63e	---	---	---	4SK4	N100E86	7a	UO	63e	Sampson
221	Modoc Glass Mt.	Long	63f	---	---	---	4SK4	N100E86	7a	UO	63f	Sampson
222	Modoc Glass Mt.	Long	63i	---	---	---	4SK4	N100E86	7a	UO	63i	Sampson
223	Modoc Glass Mt.	Long	63j	---	---	---	4SK4	N100E86	7a	UO	63j	Sampson
224	Modoc Glass Mt.	Long	64c	---	---	---	4SK4	N100E86	7a	UO	64c	Sampson
225	Modoc Glass Mt.	Long	64e	---	---	---	4SK4	N100E86	7a	UO	64e	Sampson
226	Modoc Glass Mt.	Long	65a	---	---	---	4SK4	N100E86	7b	UO	65a	Sampson
227	Modoc Glass Mt.	Long	65b	---	---	---	4SK4	N100E86	7b	UO	65b	Sampson
228	Modoc Glass Mt.	Long	65c	---	---	---	4SK4	N100E86	7b	UO	65c	Sampson
229	Modoc Glass Mt.	Long	65d	---	---	---	4SK4	N100E86	7b	UO	65d	Sampson
230	Modoc Glass Mt.	Long	65e	---	---	---	4SK4	N100E86	7b	UO	65e	Sampson
231	Modoc Glass Mt.	Long	65g	---	---	---	4SK4	N100E86	7b	UO	65g	Sampson
232	Modoc Glass Mt.	Long	65i	---	---	---	4SK4	N100E86	7b	UO	65i	Sampson
233	Modoc Glass Mt.	Long	65k	---	---	---	4SK4	N100E86	7b	UO	65k	Sampson
234	Modoc Glass Mt.	Long	66a	---	---	---	4SK4	N100E86	7b	UO	66a	Sampson
235	Modoc Glass Mt.	Long	66c	---	---	---	4SK4	N100E86	7b	UO	66c	Sampson
236	Modoc Glass Mt.	Long	66g	---	---	---	4SK4	N100E86	7b	UO	66g	Sampson
237	Modoc Glass Mt.	Long	66i	---	---	---	4SK4	N100E86	8	UO	66i	Sampson
238	Modoc Glass Mt.	Long	70b	---	---	---	4SK4	N100E86	8	UO	70b	Sampson
239	Modoc Glass Mt.	Long	70c	---	---	---	4SK4	N100E86	8	UO	70c	Sampson
240	Modoc Glass Mt.	Long	70e	---	---	---	4SK4	N100E86	8	UO	70e	Sampson
241	Modoc Glass Mt.	Long	70f	---	---	---	4SK4	N100E86	8	UO	70f	Sampson
242	Modoc Glass Mt.	Long	70g	---	---	---	4SK4	N100E86	8	UO	70g	Sampson
243	Modoc Glass Mt.	Long	70h	---	---	---	4SK4	N100E86	8	UO	70h	Sampson
244	Modoc Glass Mt.	Long	70i	---	---	---	4SK4	N100E86	8	UO	70i	Sampson
245	Modoc Glass Mt.	Long	70j	---	---	---	4SK4	N100E86	8	UO	70j	Sampson

Appendix A2 (contd.)

No.	Source	Tech.	Anal. No.	Prog. No.	Prob.	D^2	Site	Unit	Level	Coll.	Art. No.	Sub.
246	Modoc Glass Mt.	Long	70k	---	---	---	4SK4	N100E86	8	UO	70k	Sampson
247	Modoc Glass Mt.	Long	71b	---	---	---	4SK4	N100E86	8	UO	71b	Sampson
248	Modoc Glass Mt.	Long	71c	---	---	---	4SK4	N100E86	8	UO	71c	Sampson
249	Modoc Glass Mt.	Long	71d	---	---	---	4SK4	N100E86	8	UO	71d	Sampson
250	Modoc Glass Mt.	Long	71g	---	---	---	4SK4	N100E86	8	UO	71g	Sampson
251	Modoc Glass Mt.	Long	72a	---	---	---	4SK4	N100E86	9a	UO	72a	Sampson
252	Modoc Glass Mt.	Long	72b	---	---	---	4SK4	N100E86	9a	UO	72b	Sampson
253	Modoc Glass Mt.	Long	72c	---	---	---	4SK4	N100E86	9a	UO	72c	Sampson
254	Modoc Glass Mt.	Long	72d	---	---	---	4SK4	N100E86	9a	UO	72d	Sampson
255	Modoc Glass Mt.	Long	72f	---	---	---	4SK4	N100E86	9a	UO	72f	Sampson
256	Modoc Glass Mt.	Long	73b	---	---	---	4SK4	N100E86	9a	UO	73b	Sampson
257	Modoc Glass Mt.	Long	73f	---	---	---	4SK4	N100E86	9a	UO	73f	Sampson
258	Modoc Glass Mt.	Long	75a	---	---	---	4SK4	N100E86	9b	UO	75a	Sampson
259	Modoc Glass Mt.	Long	75b	---	---	---	4SK4	N100E86	9b	UO	75b	Sampson
260	Modoc Glass Mt.	Long	75c	---	---	---	4SK4	N100E86	9b	UO	75c	Sampson
261	Modoc Glass Mt.	Long	75g	---	---	---	4SK4	N100E86	9b	UO	75g	Sampson
262	Modoc Glass Mt.	Long	76a	---	---	---	4SK4	N100E86	10	UO	76a	Sampson
263	Modoc Glass Mt.	Long	76b	---	---	---	4SK4	N100E86	10	UO	76b	Sampson
264	Modoc Glass Mt.	Long	76c	---	---	---	4SK4	N100E86	10	UO	76c	Sampson
265	Modoc Glass Mt.	Long	76e	---	---	---	rSK4	N100E86	10	UO	76e	Sampson
266	Modoc Glass Mt.	Long	76h	---	---	---	4SK4	N100E86	10	UO	76h	Sampson
267	Modoc Glass Mt.	Long	76j	---	---	---	4SK4	N100E86	10	UO	76j	Sampson
268	Modoc Glass Mt.	Long	89b	---	---	---	4SK4	N84E44	2	UO	89b	Sampson
269	Modoc Glass Mt.	Long	90c	---	---	---	4SK4	N84E44	2	UO	90c	Sampson
270	Modoc Glass Mt.	Long	90d	---	---	---	4SK4	N84E44	2	UO	90d	Sampson
271	Modoc Glass Mt.	Long	131b	---	---	---	4SK4	N72E100	1b	UO	131b	Sampson
272	Modoc Glass Mt.	Long	131c	---	---	---	4SK4	N72E100	1b	UO	131c	Sampson
273	Modoc Glass Mt.	Long	131f	---	---	---	4SK4	N72E100	1b	UO	131f	Sampson

Appendix 2A (contd.)

No.	Source	Tech.	Anal. No.	Prog. No.	Prob.	D^2	Site	Unit	Level	Coll.	Art. No.	Sub.
274	Modoc Glass Mt.	Long	131h	---	---	---	4SK4	N72E100	1b	UO	131h	Sampson
275	Mono Craters	Short	1663	CC6	0.554	2.9	Mad-179	H.P. 15	120-140cm	---	---	Stickel
276	Mono Craters	Short	1556	CC6	0.959	4.7	Mad-179	H.P. 15	120-140cm	---	5	Stickel
277	Mono Craters	Short	1573	CC6	0.454	4.7	Mad-179	H.P. 15	120-140cm	---	12	Stickel
278	Mono Craters	Short	1575	CC6	0.995	26.0	Mad-179	H.P. 15	120-140cm	---	14	Stickel
279	Mono Craters	Short	1576	CC6	0.972	12.3	Mad-179	H.P. 15	120-140cm	---	1	Stickel
280	Mono Craters	Short	1578	CC6	0.607	19.5	Mad-179	H.P. 15	120-140cm	---	3	Stickel
281	Mono Craters	Short	1579	CC6	0.987	32.5	Mad-179	H.P. 15	140-160cm	---	4	Stickel;
282	Mono Craters	Short	1580	CC6	0.903	14.7	Mad-179	H.P. 15	140-160cm	---	5	Stickel
283	Mono Craters	Short	1581	CC6	0.626	11.2	Mad-179	H.P. 15	140-160cm	---	6	Stickel
284	Mono Craters	Short	1582	CC6	0.555	8.2	Mad-179	H.P. 15	140-160cm	---	7	Stickel
285	Mono Craters	Long	1619	LC1	0.499	165	Iny-382	Rose Sp.	60-64"	UCB	1-188109	JEE
286	Mono Craters	Long	1631	LC1	0.998	378	Iny-382	Rose Sp.	72-84"	UCB	1-188174	JEE
287	Mono Glass Mt.	Short	677	CC2	0.946	3.4	Mrp-105	Pit 22C	18-24"	UCLA	335-1710	JEE
288	Mono Glass Mt.	Short	680	CC2	0.803	3.8	Mrp-105	Pit 22C	18-24"	UCLA	335-1715	JEE
289	Mono Glass Mt.	Short	683	CC2	0.910	5.5	Mrp-105	Pit 22C	18-24"	UCLA	335-1718	JEE
290	Mono Glass Mt.	Short	684	CC2	0.914	2.2	Mrp-105	Pit 22C	18-24"	UCLA	335-1720	JEE
291	Mono Glass Mt.	Short	685	CC2	0.963	4.0	Mrp-105	Pit 22C	18-24"	UCLA	335-1721	JEE
292	Mono Glass Mt.	Short	688	CC2	0.935	3.2	Mrp-105	Pit 22C	36-42"	UCLA	335-1751	JEE
293	Mono Glass Mt.	Short	689	CC2	0.759	6.2	Mrp-105	Pit 22C	36-42"	UCLA	335-1752	JEE
294	Mono Glass Mt.	Short	760	CC2	0.914	2.4	Sjo-68	Bur. 46	25"	UCB	1-173319	Bennyhoff
295	Mono Glass Mt.	Short	765	CC2	0.929	4.4	Ala-309	Tr. 3/HP3	27'7"	UCB	1-26007	Bennyhoff
296	Mono Glass Mt.	Short	589	S2	0.886	11.6	LAn-324	No/W1	0-15cm	---	W66-1602	Singer
297	Mono Glass Mt.	Long	1666	LC1	1.0	22	Cal 99	Pit 3	16"	UCB	1-139122	JEE
298	Mt. Konocti	XRF	995	---	---	---	Lak-380	Bur. 4	---	SDA	M-4	R. King
299	Mt. Konocti	XRF	996	---	---	---	Lak-380	Bur. 4	---	SDA	M-5	R. King
300	Mt. Konocti	XRF	998	---	---	---	Lak-380	Bur. 4	---	SDA	M-7	R. King

Appendix 2A (contd.)

No.	Source	Tech.	Anal. No.	Prog. No.	Prob.	D^2	Site	Unit	Level	Coll.	Art. No.	Sub.
301	Mt. Konocti	XRF	---	---	---	---	Lak-380	Bur. 1	---	SDA	1	R. King
302	Mt. Konocti	XRF	---	---	---	---	Lak-380	Bur. 1	---	SDA	2	R. King
303	Mt. Konocti	XRF	---	---	---	---	Lak-380	Bur. 1	---	SDA	3	R. King
304	Mt. Konocti	XRF	---	---	---	---	Lak-380	Bur. 1	---	SDA	4	R. King
305	Mt. Konocti	XRF	---	---	---	---	Lak-380	Bur. 1	---	SDA	5	R. King
306	Mt. Konocti	XRF	---	---	---	---	Lak-380	Bur. 1	---	SDA	6	R. King
307	Mt. Konocti	XRF	---	---	---	---	Lak-380	Bur. 1	---	SDA	7	R. King
308	Mt. Konocti	XRF	---	---	---	---	Lak-380	Bur. 1	---	SDA	1	R. King
309	Mt. Konocti	XRF	---	---	---	---	Lak-380	Bur. 1	---	SDA	2	R. King
310	Mt. Konocti	XRF	---	---	---	---	Lak-380	Bur. 1	---	SDA	3	R. King
311	Mt. Konocti	XRF	---	---	---	---	Lak-380	Bur. 1	---	SDA	4	R. King
312	Mt. Konocti	XRF	---	---	---	---	Lak-380	Bur. 1	---	SDA	6	R. King
313	Mt. Konocti	XRF	---	---	---	---	Lak-380	Bur. 1	---	SDA	7	R. King
314	Mt. Konocti	Short	769	CC2	0.854	24.4	Sjo-68	Crem. 1	47"	UCB	1-73243c	JEE
315	Mt. Konocti	Short	772	CC2	0.971	52.4	Sjo-68	Crem. 1	47"	UCB	1-73243g	JEE
316	Napa Glass Mt.	Short	141	CC2	1.0	45.6	CCo-138	Bur. 22	37"	UCB	1-49241A	Bennyhoff
317	Napa Glass Mt.	Short	142	CC2	1.0	53.7	CCo-138	Bur. 22	37"	UCB	1-49241C	Bennyhoff
318	Napa Glass Mt.	Short	143	CC2	1.0	53.5	CCo-138	Bur. 22	37"	UCB	1-49241D	Bennyhoff
319	Napa Glass Mt.	Short	144	CC2	1.0	46.2	CCo-138	Bur. 22	37"	UCB	1-49241E	Bennyhoff
320	Napa Glass Mt.	Short	145	CC2	1.0	57.8	CCo-138	Bur. 22	37"	UCB	1-49241F	Bennyhoff
321	Napa Glass Mt.	Short	146	CC2	1.0	48.8	CCo-138	Bur. 22	37"	UCB	1-49241G	Bennyhoff
322	Napa Glass Mt.	Short	147	CC2	1.0	40.4	CCo-138	Bur. 22	37"	UCB	1-49241H	Bennyhoff
323	Napa Glass Mt.	Short	148	CC2	1.0	38.3	CCo-138	Bur. 22	37"	UCB	1-49241H	Bennyhoff
324	Napa Glass Mt.	Short	149	CC2	1.0	39.4	CCo-138	Bur. 22	37"	UCB	1-49241J	Bennyhoff
325	Napa Glass Mt.	Short	150	CC2	1.0	28.2	CCo-138	Bur. 22	37"	UCB	1-49241K	Bennyhoff
326	Napa Glass Mt.	Short	151	CC2	1.0	31.1	CCo-138	Bur. 22	37"	UCB	1-49241L	Bennyhoff
327	Napa Glass Mt.	Short	152	CC2	1.0	33.6	CCo-138	Bur. 22	37"	UCB	1-49241M	Bennyhoff

Appendix 2A (contd.)

No.	Source	Tech.	Anal. No.	Prog. No.	Prob.	D^2	Site	Unit	Level	Coll.	Art. No.	Sub.
328	Napa Glass Mt.	Short	153	CC2	1.0	46.7	CCo-138	Bur. 22	37"	UCB	1-49241N	Bennyhoff
329	Napa Glass Mt.	Short	154	CC2	1.0	38.0	CCo-138	Bur. 22	37"	UCB	1-492410	Bennyhoff
330	Napa Glass Mt.	Short	155	CC2	1.0	50.9	CCo-138	Bur. 22	37"	UCB	1-49241P	Bennyhoff
331	Napa Glass Mt.	Short	156	CC2	1.0	33.5	CCo-138	Bur. 22	37"	UCB	1-492410	Bennyhoff
332	Napa Glass Mt.	Short	157	CC2	1.0	40.2	CCo-138	Bur. 22	37"	UCB	1-49241R	Bennyhoff
333	Napa Glass Mt.	Short	158	CC2	1.0	41.1	CCo-138	Bur. 22	37"	UCB	1-49241S	Bennyhoff
334	Napa Glass Mt.	Short	159	CC2	1.0	41.3	CCo-138	Bur. 22	37"	UCB	1-49241T	Bennyhoff
335	Napa Glass Mt.	Short	757	CC2	0.999	1.3	CCo-138	Bur. 9	20"	UCB	1-39189	Bennyhoff
336	Napa Glass Mt.	Short	758	CC2	1.0	1.7	CCo-138	Bur. 9	20"	UCB	1-39190	Bennyhoff
337	Napa Glass Mt.	Short	759	CC2	1.0	1.1	CCo-138	Bur. 9	20"	UCB	1-39192	Bennyhoff
338	Napa Glass Mt.	Short	762	CC2	1.0	3.0	SJo-68	Crem. 3	50"	UCB	1-73364	Bennyhoff
339	Napa Glass Mt.	Short	763	CC2	1.0	0.2	Ala-309	Tr. 1	26' 4"	UCB	1-25999	Bennyhoff
340	Napa Glass Mt.	Short	764	CC2	1.0	0.9	Ala-309	Bur. 3801	Tr. 2	UCB	1-26004	Bennyhoff
341	Napa Glass Mt.	Short	766	CC2	0.980	1.8	SJo-68	Crem. 1	47"	UCB	1-73234	Bennyhoff
342	Napa Glass Mt.	Short	773	CC2	1.0	0.5	Sol-236	Bur. 12	78"	UCB	1-22140	Bennyhoff
343	Napa Glass Mt.	Short	774	CC2	1.0	2.7	Sol-236	Bur. 12	78"	UCB	1-22147A	Bennyhoff
344	Napa Glass Mt.	Short	775	CC2	1.0	0.7	Sol-236	Bur. 12	78"	UCB	1-22147B	Bennyhoff
345	Napa Glass Mt.	Short	776	CC2	0.940	1.6	Sol-236	Bur. 12	78"	UCB	1-22138	Bennyhoff
346	Napa Glass Mt.	Short	777	CC2	1.0	2.0	Sol-236	Bur. 12	78"	UCB	1-22148A	Bennyhoff
347	Napa Glass Mt.	Short	778	CC2	0.999	0.5	Sol-236	Bur. 12	78"	UCB	1-22148B	Bennyhoff
348	Napa Glass Mt.	Short	779	CC2	1.0	2.2	Sol-236	Bur. 12	78"	UCB	1-22147C	Bennyhoff
349	Napa Glass Mt.	Short	780	CC2	1.0	3.8	CCo-138	Bur. 8	20"	UCB	1-39167	Bennyhoff
350	Napa Glass Mt.	Short	781	CC2	1.0	2.6	CCo-138	Bur. 8	20"	UCB	1-39168	Bennyhoff
351	Napa Glass Mt.	Short	782	CC2	1.0	2.7	CCo-138	Bur. 8	20"	UCB	1-39169	Bennyhoff
352	Napa Glass Mt.	Short	783	CC2	1.0	1.6	CCo-138	Bur. 8	20"	UCB	1-39170	Bennyhoff
353	Napa Glass Mt.	Short	784	CC2	1.0	0.5	CCo-138	Bur. 8	20"	UCB	1-39165	Bennyhoff

Appendix 2A (contd.)

No.	Source	Tech.	Anal. No.	Prog. No.	Prob.	D^2	Site	Unit	Level	Coll.	Art. No.	Sub.
354	Napa Glass Mt.	Short	785	CC2	1.0	0.2	CCo-138	Bur. 8	20"	UCB	1-39166B	Bennyhoff
355	Napa Glass Mt.	Short	786	CC2	0.976	1.0	CCo-138	Bur. 8	20"	UCB	1-39166A	Bennyhoff
356	Napa Glass Mt.	Short	795	CC2	1.0	0.2	CCo-138	Bur. 22	37"	UCB	1-49207A	Bennyhoff
357	Napa Glass Mt.	Short	796	CC2	0.999	0.0	Sac-107	Bur. C8	52"	UCB	1-46330	Bennyhoff
358	Napa Glass Mt.	Short	797	CC2	0.993	0.7	Sac-107	Bur. C8	52"	UCB	1-46339	Bennyhoff
359	Napa Glass Mt.	Short	810	CC4	0.986	27.7	Son-518	H.F.	Feature 1	SDA	73-9-003	Upson
360	Napa Glass Mt.	Short	815	CC4	0.998	45.5	Son-518	H.F.	Feature 1	SDA	73-9-005	Upson
361	Napa Glass Mt.	Short	869	CC4	1.0	4.3	CCo-30	I-28	18-24"	SDA	1	Frederickson
362	Napa Glass Mt.	Short	870	CC4	1.0	25.9	CCo-30	I-28	18-24"	SDA	2	Frederickson
363	Napa Glass Mt.	Short	871	CC4	1.0	2.8	CCo-30	I-28	18-24"	SDA	3	Frederickson
364	Napa Glass Mt.	Short	872	CC4	1.0	29.9	CCo-30	I-28	18-24"	SDA	4	Frederickson
365	Napa Glass Mt.	Short	873	CC4	0.909	3.9	CCo-30	I-28	18-24"	SDA	5	Frederickson
366	Napa Glass Mt.	Short	874	CC4	1.0	6.5	CCo-20	I-28	18-24"	SDA	6	Frederickson
367	Napa Glass Mt.	Short	875	CC4	1.0	5.6	CCo-30	I-28	18-24"	SDA	7	Frederickson
368	Napa Glass Mt.	Short	876	CC4	1.0	3.6	CCo-30	I-28	18-24"	SDA	8	Frederickson
369	Napa Glass Mt.	Short	877	CC4	1.0	1.2	CCo-30	I-28	18-24"	SDA	9	Frederickson
370	Napa Glass Mt.	Short	878	CC4	1.0	4.2	CCo-30	I-28	18-24"	SDA	10	Frederickson
371	Napa Glass Mt.	Short	879	CC4	1.0	2.4	CCo-30	I-28	18-24"	SDA	11	Frederickson
372	Napa Glass Mt.	Short	880	CC4	1.0	4.6	CCo-30	I-28	18-24"	SDA	12	Frederickson
373	Napa Glass Mt.	Short	881	CC4	1.0	6.0	CCo-30	I-28	36-42"	SDA	1	Frederickson
374	Napa Glass Mt.	Short	882	CC4	1.0	28.8	CCo-30	I-28	36-42"	SDA	2	Frederickson
375	Napa Glass Mt.	Short	884	CC4	1.0	1.2	CCo-30	I-28	36-42"	SDA	4	Frederickson
376	Napa Glass Mt.	Short	885	CC4	1.0	2.4	CCo-30	I-28	36-42"	SDA	5	Frederickson
377	Napa Glass Mt.	Short	886	CC4	1.0	2.3	CCo-30	I-28	36-42"	SDA	6	Frederickson
378	Napa Glass Mt.	Short	887	CC4	1.0	5.0	CCo-30	I-28	36-42"	SDA	7	Frederickson
379	Napa Glass Mt.	Short	888	CC4	1.0	9.9	CCo-30	I-28	36-42"	SDA	8	Frederickson
380	Napa Glass Mt.	Short	889	CC4	1.0	0.8	CCo-30	I-28	36-42"	SDA	9	Frederickson

Appendix 2A (contd.)

No.	Source	Tech.	Anal. No.	Prog. No.	Prob.	D^2	Site	Unit	Level	Coll.	Art. No.	Sub.
381	Napa Glass Mt.	Short	890	CC4	1.0	6.2	CCo-30	I-28	36-42"	SDA	10	Frederickson
382	Napa Glass Mt.	Short	891	CC4	1.0	5.1	CCo-30	I-28	36-42"	SDA	11	Frederickson
383	Napa Glass Mt.	Short	892	CC4	1.0	3.8	CCo-30	I-28	36-42"	SDA	12	Frederickson
384	Napa Glass Mt.	Short	893	CC4	1.0	3.6	CCo-30	Q-40	18-24"	SDA	1	Frederickson
385	Napa Glass Mt.	Short	895	CC4	1.0	0.3	CCo-30	Q-40	18-24"	SDA	3	Frederickson
386	Napa Glass Mt.	Short	896	CC4	1.0	1.5	CCo-30	Q-40	18-24"	SDA	4	Frederickson
387	Napa Glass Mt.	Short	897	CC4	1.0	6.7	CCo-30	Q-40	18-24"	SDA	5	Frederickson
388	Napa Glass Mt.	Short	899	CC4	1.0	2.1	CCo-30	Q-40	18-24"	SDA	7	Frederickson
389	Napa Glass Mt.	Short	900	CC4	1.0	3.2	CCo-30	Q-40	18-24"	SDA	8	Frederickson
390	Napa Glass Mt.	Short	901	CC4	1.0	2.6	CCo-30	Q-40	18-24"	SDA	9	Frederickson
391	Napa Glass Mt.	Short	902	CC4	1.0	9.9	CCo-30	Q-40	18-24"	SDA	10	Frederickson
392	Napa Glass Mt.	Short	903	CC4	1.0	25.4	CCo-30	Q-40	18-24"	SDA	11	Frederickson
393	Napa Glass Mt.	Short	904	CC4	1.0	5.3	CCo-30	Q-40	18-24"	SDA	12	Frederickson
394	Napa Glass Mt.	Short	905	CC4	0.911	5.3	CCo-30	Q-40	36-42"	SDA	1	Frederickson
395	Napa Glass Mt.	Short	907	CC4	1.0	1.6	CCo-30	Q-40	36-42"	SDA	3	Frederickson
396	Napa Glass Mt.	Short	908	CC4	1.0	2.8	CCo-30	Q-40	36-42"	SDA	4	Frederickson
397	Napa Glass Mt.	Short	909	CC4	1.0	1.9	CCo-30	Q-40	36-42"	SDA	5	Frederickson
398	Napa Glass Mt.	Short	911	CC4	1.0	2.0	CCo-30	Q-40	36-42"	SDA	7	Frederickson
399	Napa Glass Mt.	Short	912	CC4	1.0	1.7	CCo-30	Q-40	36-42"	SDA	8	Frederickson
400	Napa Glass Mt.	Short	913	CC4	0.985	1.4	CCo-30	Q-40	36-42"	SDA	9	Frederickson
401	Napa Glass Mt.	Short	914	CC4	1.0	1.2	CCo-30	Q-40	36-42"	SDA	10	Frederickson
402	Napa Glass Mt.	Short	915	CC4	1.0	2.8	CCo-30	Q-40	36-42"	SDA	11	Frederickson
403	Napa Glass Mt.	Short	916	CC4	1.0	0.6	CCo-30	Q-40	36-42"	SDA	12	Frederickson
404	Napa Glass Mt.	Short	1195	CC4	1.0	2.6	Mrn-152	Bur. 5	90 cm	---	---	Clewlow
405	Napa Glass Mt.	Short	1196	CC4	1.0	6.1	Mrn-152	Bur. 5	90 cm	---	---	Clewlow
406	Napa Glass Mt.	Short	1198	CC4	1.0	4.3	Mrn-152	Bur. 5	90 cm	---	---	Clewlow
407	Napa Glass Mt.	Short	1199	CC4	1.0	7.0	Mrn-152	Bur. 5	90 cm	---	---	Clewlow

Appendix 2A (contd.)

No.	Source	Tech.	Anal. No.	Prog. No.	Prob.	D^2	Site	Unit	Level	Coll.	Art. No.	Sub.
408	Napa Glass Mt.	Short	1200	CC4	1.0	7.4	Mrn-152	Bur. 5	90 cm	---	---	Clewlow
409	Napa Glass Mt.	Short	1201	CC4	1.0	1.4	Mrn-152	Bur. 4	85-90 cm	---	---	Clewlow
410	Napa Glass Mt.	Short	1203	CC4	1.0	2.6	Mrn-152	Bur. 4	85-90 cm	---	---	Clewlow
411	Napa Glass Mt.	Short	1204	CC4	0.998	0.3	Mrn-152	Bur. 4	85-90 cm	---	---	Clewlow
412	(W)	Long	1656	LC1	0.454	17	Nev-15	16 SW20	12-24"	UCB	1-173134-2	JEE
413	(W)	Long	1660	LC1	0.655	23	Ala-307	H-4	150"	UCB	1-122881-1	JEE
414	(W)	Long	1678	LC1	0.788	19	CCo-138		50"	UCB	1-163947-4	JEE
415	(W)	Long	1682	LC1	0.753	10	CCo-138		50"	UCB	1-163947-8	JEE
416	(W)	Long	1683	LC1	0.749	12	CCo-138	IG-55	50"	UCB	1-163947-9	JEE
417	(S)	Long	1674	LC1	0.928	25	CCo-138		50"	UCB	1-226486	JEE
418	(S)	Long	1675	LC1	0.356	20	CCo-138		50"	UCB	1-163947-1	JEE
419	(S)	Long	1680	LC1	0.354	17	CCo-138		50"	UCB	1-163947-6	JEE
420	(S)	Long	1686	LC1	0.570	14	CCo-138		50"	UCB	1-163947-12	JEE
421	(E. Dago)	Long	1652	LC1	1.0	68	Sol-2		30"	UCB	1-80106	JEE
422	(E. Dago)	Long	1653	LC1	0.942	13	Sol-2		30"	UCB	1-80107	JEE
423	(E. Dago)	Long	1654	LC1	0.377	18	Sol-2		30"	UCB	1-80108	JEE
424	(E. Dago)	Long	1661	LC1	0.613	12	Ala-307	H-4	150"	UCB	1-122881-2	JEE
425	(E. Dago)	Long	1662	LC1	0.319	14	Ala-307	H-4	150"	UCB	1-122882	JEE
426	Napa Glass Mt.	Long	1670	LC1	0.436	13	Cal-99	H-4	24"	UCB	1-139117	JEE
427	(E. Dago)	Long	1677	LC1	0.575	14	CCo-138	1G-S5	50"	UCB	1-163947-3	JEE
428	(E. Dago)	Long	1676	LC1			CCo-138	1G-S5	50"	UCB	1-163947-2	JEE
429	(E. Dago)	Long	1679	LC1	0.840	18	CCo-138	1G-S5	50"	UCB	1-163947-5	JEE
430	(E. Dago)	Long	1681	LC1	0.469	17	CCo-138	1G-S5	50"	UCB	1-163947-7	JEE
431	(E. Dago)	Long	1685	LC1	0.895	2	CCo-138	1G-S5	50"	UCB	1-163947-11	JEE
432	(E. Dago)	Long	1687	LC1	0.595	10	CCo-138	1G-S5	50"	UCB	1-163947-13	JEE
433	(E. Dago)	Long	1701	LC1	0.432	16	CCo-138	1G-S5	50"	UCB	1-163947-27	JEE
434	(E. Dago)	Long	1703	LC1	0.631	16	CCo-138	1G-S5	50"	UCB	1-163947-29	JEE

Appendix 2A (contd.)

No.	Source	Tech.	Anal. No.	Prog. No.	Prob.	D²	Site	Unit	Level	Coll.	Art. No.	Sub.
435	(E. Dago)	Long	1704	LC1	0.999	39	CCo-138	1G-S5	50"	UCB	1-163947-30	JEE
436	(E. Dago)	Long	1751	LC1	0.339	16	E1D-44	Tr.A.Sq 2	12-24"	UCB	1-197564	JEE
437	(E. Dago)	Long	1684	LC1	0.465	10	CCo-138	1G-S5	50"	UCB	1-163947-10	JEE
438	Obsidian Butte	Short	645	S2	1.0	1.8	Riv-463	Area 2	60-70 cm	---	G1	Wilke
439	Obsidian Butte	Short	646	S2	1.0	13.7	Riv-463	Area 2	60-70 cm	---	G2	Wilke
440	Obsidian Butte	Short	647	S2	1.0	26.8	Riv-463	Area 2	60-70 cm	---	G3	Wilke
441	Obsidian Butte	Short	632	S2	1.0	.	Riv-463	Area 2	30-40 cm	---	D7	Wilke
442	Mt. Hicks	Short	682	CC2	0.863	36.1	Mrp-105	Pit 22C	18-24"	UCLA	335-1717	JEE
443	Mt. Hicks	Short	690	CC2	1.0	39.9	Mrp-105	Pit 22C	36-42"	UCLA	335-1753	JEE
444	Mt. Hicks	Short	692	CC2	1.0	41.4	Mrp-105	Pit 22C	36-42"	UCLA	335-1756	JEE
445	Mt. Hicks	Short	694	CC2	1.0	36.2	Mrp-105	Pit 22C	36-42"	UCLA	335-1759	JEE
446	Mt. Hicks	Long	1664	LC1	0.999	10	Cal-99	Pit 3	21"	UCB	1-139118	JEE
447	Mt. Hicks	Long	1665	LC1	1.0	9	Cal-99	Pit 3	14"	UCB	1-139121	JEE
448	Pine Grove Hills	Short	676	CC2	0.999	---	Mrp-105	Pit 22C	18-24"	UCLA	335-1709	JEE
449	Pine Grove Hills	Short	768	CC2	0.922	---	SJo-68	Crem. 1	47"	UCB	1-173243B	Bennyhoff
450	Pine Grove Hills	Short	767	CC2	0.935	---	SJo-68	Crem. 1	47"	UCB	1-73241	Bennyhoff

Appendix 2B

ASSOCIATED DATE AND HYDRATION MEASUREMENT DATA

No.	Source	Lab#	C¹⁴ Date	Material	Cal. Date	Reference	Assoc.	Lab#	Hydra	Mean	Analyst
1	Annadel	UCLA 1793C	365±50	Charcoal	1380 AD	Berger & Libby(nd)	Direct	3731	2.5	2.50	Bennett
2	"	UCLA 1793D	465±50	"	1450 AD	"	"	3743	2.2	2.20	"
3	"	UCLA 1891B	3270±70	Collagen	1625 BC	"	"	4038	2.5	2.65	"
4	"	"	"	"	"	"	"	4041	2.8		"
5	"	UCLA 1891A	3050±130	"	1300 BC	"	"	4046	2.1	2.70	"
6	"	"	"	"	"	"	"	4049	3.3		"
7	"	M-127B	3700±300	Charcoal	2140 BC	Crane 1956	Strat	3143	4.3	4.30	"
8	"	UCLA 276	950±70	"	1025 AD	Fergusson & Libby '64	Direct	3077	9.2	6.18	"
9	"	"	"	"	"	"	"	3078	4.6		"
10	"	"	"	"	"	"	"	3081	4.6		"
11	"	"	"	"	"	"	"	3086	6.3		"
12	Bodie Hills	UCLA 277	1560±30	"	360 AD	Fergusson & Libby '64	"	3093	3.7	3.70	"
13	"	I-2750B	2585±100	Collagen	790-8252	Ragir '72	"	3144	4.8	4.80	"
14	"	M-647	4350±250	Bone	2900 BC	Crane & Griffin '58	Strat	3133	5.5	5.50	"
15	"	UCLA 1794C	115±45	Charcoal	1680 AD	Berger & Libby n.d.	Direct	4105	2.1	2.03	"
16	"	"	"	"	"	"	"	4109	1.5		"
17	"	"	"	"	"	"	"	4110	2.5		"
18	"	M-123	2880±300	"	1120-1200BC	Crane 1956	"	4818	3.9	3.90	"
19	"	UCLA 1952A	1620±400	Collagen	400 AD	Berger & Libby n.d.	"	4822	2.4	2.40	"
20	"	UCLA 1952B	1230±100	"	700 AD	"	"	4826	2.3	2.35	"

No.	Source	Lab#	C14 Date	Material	Cal. Date	Reference	Assoc.	Lab#	Hydra	Mean	Analyst
21	Bodie Hills	UCLA 1952B	1200±100	Collagen	700 AD	Berger & Libby n.d.	Direct	4828	2.2	2.35	Bennett
22	"	"	"	"	"	"	"	4830	2.3	"	"
23	"	"	"	"	"	"	"	4832	2.6	"	"
24	Borax Lake	UCLA 1794C	115±45	Charcoal	1680 AD	"	"	4108	2.3	2.30	VB
25	"	"	"	"	"	"	"	4107	2.1	"	"
26	"	"	"	"	"	"	"	4103	2.5	"	"
27	"	UCLA 1793D	465±50	"	1405 AD	Ericson & Berger '74	"	3747	2.6	2.60	"
28	"	UCLA 1795A	10260±340	Collagen	-	"	"	3754	6.4	6.28	"
29	"	"	"	"	-	"	"	3755	5.7/12.4	"	"
30	"	"	"	"	-	"	"	3756	6.2	"	"
31	"	"	"	"	-	"	"	3759	6.8	"	"
32	"	UCLA 1795C	7750±400	"	-	Berger & Libby n.d.	"	4100	7.4/16.0	7.40	"
33	"	UCLA 1951A	1950±60	Charcoal	60 AD	"	"	4814	4.4	4.40	"
34	"	UCLA 1951B	275±40	"	1510-1640AD	"	Strat.	4817	3.5	3.50	"
35	"	M-647	4350±250	Bone	2900 BC	Crane & Griffin '58	"	3134	4.1	4.10	"
36	"	"	3690±130	"	2120 BC	"	Direct	1815	3.9	3.77	FJF/SDA
37	"	"	"	"	"	"	"	1816	3.6	"	"
38	"	"	"	"	"	"	"	1817	3.7	"	"
39	"	"	"	"	"	"	"	1818	3.8	"	"
40	"	"	"	"	"	"	"	1819	3.7	"	"
41	"	"	"	"	"	"	"	1820	3.7	"	"
42	"	"	"	"	"	"	"	1821	4.3	"	"
43	"	"	"	"	"	"	"	1822	3.9	"	"
44	"	"	"	"	"	"	"	1823	3.6	"	"
45	"	"	"	"	"	"	"	1824	4.2	"	"
46	"	"	"	"	"	"	"	1825	3.1	"	"
47	"	"	2100±150	"	130 BC	"	"	2228	2.6	3.21	"
48	"	"	"	"	"	"	"	2229	2.8	"	"

207

No.	Source	Lab#	C14 Date	Material	Cal. Date	Reference	Assoc.	Lab#	Hydra	Mean	Analyst
49	Borax Lake	M-647	2100±150	Bone	130 BC	Crane & Griffin '58	Direct	2230	3.6	3.21	FJF/SDA
50	"	"	"	"	"	"	"	2231	4.1	"	"
51	"	"	"	"	"	"	"	2234	2.2/3.3	"	"
52	"	"	"	"	"	"	"	2235	4.3	"	"
53	"	"	"	"	"	"	"	2236	3.1	"	"
54	"	"	"	"	"	"	"	2237	3.1	"	"
55	"	0	"	0	"	0	"	2238	3.1	"	"
56	Buck Mt	none	none	-	-	-	-	54c	2.1	1.75	Johnson
57	"	"	"	-	-	-	-	54d	1.4	"	"
58	"	Gak-1831	2180±80	Charcoal	180 BC	Johnson '69	Direct?	58b	3.1	2.25	"
59	"	"	"	"	"	"	"	58d	1.4	"	"
60	"	Gak-1832	2340±100	"	450 BC	"	"	59b	3.1	3.10	"
61	"	Gak-1833	2180±90	"	180 BC	"	"	62a	3.3	3.10	"
62	"	"	"	"	"	"	"	62j	2.9	3.10	"
63	"	Gak-1834	3470±80	"	1790-2010 BC	"	"	64h	3.4	3.40	"
64	"	Gak-1835	3450±90	"	1710-1800-2010 BC	"	"	65f	3.2	3.20	"
65	"	Gak-1836	4260±100	"	2970 BC	"	"	71f	4.1	4.10	"
66	"	Gak-1837	4750±110	"	3525-3660 BC	"	"	72i	4.6	4.00	"
67	"	"	"	"	"	"	"	73d	3.5	"	"
68	"	"	"	"	"	"	"	73e	3.9	"	"
69	"	Gak-1838	4030±90	"	2970 BC	"	"	74g	5.0	4.20	"
70	"	Gak-1839	4500±110	"	"	"	"	74h	4.1	"	"

No.	Source	Lab#	C14 Date	Material	Cal. Date	Reference	Assoc.	Lab#	Hydra	Mean	Analyst
71	Buck Mt	Gak-1838 Gak-1839	4030±90 4500±100	Charcoal	2970 BC	Johnson'69	Direct?	75h	3.5	4.20	Johnson
72	"	Gak-1840	5750±130	"	4600 BC	"	"	76d	4.6	4.55	"
73	"	"	"	"	"	"	"	76f	4.5	4.5	"
74	"	Gak-2424	6160±130	"	5125 BC	"	Loose	89c	4.5	4.56	"
75	"	"	"	"	"	"	"	90b	4.5		"
76	"	"	"	"	"	"	"	90e	4.7		"
77	Casa Diablo	UCLA 1724B	430±75	"	1425 AD	Berger & Libby n.d.	Direct	2600	1.07	1.07	Michels
78	"	UCLA 1724E	445±100	"	"	"	"	2493	2.44	3.19	"
79	"	"	"	"	"	"	"	2503	3.94		"
80	"	UCLA 1724D	170±55	"	1660 AD	"	"	2402	1.59	1.59	"
81	"	UCLA 1724F	595±55	"	1330 AD	"	"	2740	3.66	3.66	"
82	"	UCLA 1724A	660±60	"	1280 AD	"	"	2597	3.42	3.30	Bennett
83	"	"	"	"	"	"	"	2389	3.17		"
84	"	UCLA 278	2040±100	"	100 BC	Fergusson & Libby 1964	"	3095	3.3	3.30	"
85	"	UCLA 277	1580±80	"	360 AD	"	"	3091	NBS/5.2	5.20	"
86	"	UCLA 1815	870±80	"	1100-1210 AD	Berger & Libby n.d.	Strat.	3167	2.9/0.4	2.90	"
87	"	UCLA 1886	1245±60	Collagen	785 AD	"	Direct	3843	5.4	5.40	"
88	"	UCLA 1863	1185±80	"	870 AD	"	Strat.	3844	5.7	5.70	"
89	"	UCLA 1920A	430±110	Charcoal	1440 AD	"	Direct	4414	7.0	6.76	"
90	"	"	"	"	"	"	"	4416	7.6		"
91	"	"	"	"	"	"	"	4419	8.9		"
92	"	"	"	"	"	"	"	4420	4.4		"

No.	Source	Lab#	C14 Date	Material	Cal. Date	Reference	Assoc.	Lab#	Hydra	Mean	Analyst
93	Casa Diablo	UCLA 1920A	430±110	Charcoal	1440 AD	Berger & Libby n.d.	Direct	4421	7.3	6.76	Bennett
94	"	"	"	"	"	"	"	4422	4.4	"	"
95	"	"	"	"	"	"	"	4423	7.5	"	"
96	"	"	"	"	"	"	"	4424	NBS	"	"
97	"	"	"	"	"	"	"	4426	7.0	"	"
98	"	UCLA 1920B	1000±300	"	1020 AD	"	"	4429	6.2/2.8	6.13	"
99	"	"	"	"	"	"	"	4435	NBS	"	"
100	"	"	"	"	"	"	"	4436	6.0	"	"
101	"	"	"	"	"	"	"	4437	6.5	"	"
102	"	"	"	"	"	"	"	4438	5.8	"	"
103	"	UCLA 1920A	430±110	"	1440 AD	"	"	4417	6.4	6.40	"
104	"Sec29	UCLA 1951A	1950±60	"	60 AD	"	"	4816	4.1	4.10	"
105	" "	UCLA 1952B	1200±100	Collagen	700 AD	"	"	4827	2.6	2.85	"
106	" "	"	"	"	"	"	"	4831	3.1	"	"
107	" "	UCLA 1954	280±55	Charcoal	1500-1640 AD	"	"	4864	4.7	4.66	"
108	"Sec35	"	"	"	"	"	"	4865	2.8	"	"
109	"	"	"	"	"	"	"	4866	NBS	"	"
110	"	"	"	"	"	"	"	4867	3.4	"	"
111	"	"	"	"	"	"	"	4868	3.7	"	"
112	"	"	"	"	"	"	"	4869	5.0	"	"
113	"Sec22	"	"	"	"	"	"	4873	4.6	"	"
114	"Sec35	"	"	"	"	"	"	4874	4.9	"	"
115	"Sec22	"	"	"	"	"	"	4875	3.9	"	"

No.	Source	Lab#	C14 Date	Material	Cal. Date	Reference	Assoc.	Lab#	Hydra	Mean	Analyst
116	Casa Diablo Sec.22	UCLA 1954	280±55	Charcoal	1500-1640 AD	Berger & Libby n.d.	Direct	4876	8.8	4.66	Bennett
117	" Sec.29	"	"	"	"	"	"	4877	3.6	"	"
118	" " 35	"	"	"	"	"	"	4878	4.9	"	"
119	" " 29	"	"	"	"	"	"	4879	6.0	"	"
120	" " 35	"	"	"	"	"	"	4880	5.6	"	"
121	" " "	"	"	"	"	"	"	4881	8.8	"	"
122	" " 29	"	"	"	"	"	"	4882	2.2	"	"
123	" " "	"	"	"	"	"	"	4883	2.8	"	"
124	" " 35	"	"	"	"	"	"	4884	5.7	"	"
125	" " "	"	"	"	"	"	"	4885	3.4	"	"
126	" " "	"	"	"	"	"	"	4886	2.8	"	"
127	" " "	"	"	"	"	"	"	4887	5.0	"	"
128	" " "	"	"	"	"	"	"	4888	5.2	"	"
129	" " "	"	"	"	"	"	"	4889	4.7	"	"
130	Coso	UCLA 1771B	2040±350	Collagen	130 BC	"	Strata		11.3	4.95	Findlow
131	"	"	"	"	"	"	"		4.4	"	"
132	"	"	"	"	"	"	"		7.0	"	"
133	"	"	"	"	"	"	"		4.1	"	"
134	"	"	"	"	"	"	"		4.3	"	"
135	"	UCLA 1771A	1720±50	Shell	280 AD	"	"		1.5	2.00	"
136	"	"	"	"	"	"	"		2.5	"	"
137	"	UCLA 1816	215±60	Charcoal	1670 AD	"	"	3158	?/0.7	0.70	Bennett

No.	Source	Lab#	C14 Date	Material	Cal. Date	Reference	Assoc.	Lab#	Hydra	Mean	Analyst
138	Coso	UCLA 1886	1245±60	Collagen	785 AD	Berger & Libby n.d.	Strata	3855	4.3	4.23	Bennett
139	"	"	"	"	"	"	"	3896	5.8	"	"
140	"	"	"	"	"	"	"	3897	4.6	"	"
141	"	"	"	"	"	"	"	3898	2.2	"	"
142	"	TX-1195	1420±70	Charcoal	675 AD	Valastro et al '75	Direct	1483	6.6	6.33	Aiello
143	"	"	"	"	"	"	"	1484	6.4	"	"
144	"	"	"	"	"	"	"	1485	6.8	"	"
145	"	"	"	"	"	"	"	1486	6.6	"	"
146	"	"	"	"	"	"	"	1487	6.8	"	"
147	"	"	"	"	"	"	"	1480	4.8	"	"
148	"	UCLA 1093A	2240±145	"	130-370 BC	Clewlow et al '70	"	4770	7.2	8.03	Bennett
149	"	"	"	"	"	"	"	4771	10.6	"	"
150	"	"	"	"	"	"	"	4772	9.8	"	"
151	"	"	"	"	"	"	"	4773	6.3	"	"
152	"	"	"	"	"	"	"	4774	6.9	"	"
153	"	"	"	"	"	"	"	4775	7.6	"	"
154	"	"	"	"	"	"	"	4776	8.7	"	"
155	"	"	"	"	"	"	"	4777	7.1	"	"
156	"	UCLA 1093B	2900±80	"	1110-1210BC	"	"	4780	5.9	8.04	"
157	"	"	"	"	"	"	"	4782	16.5	"	"
158	"	"	"	"	"	"	"	4783	8.1	"	"
159	"	"	"	"	"	"	"	4784	7.8	"	"
160	"	"	"	"	"	"	"	4785	8.2	"	"
161	"	"	"	"	"	"	"	4786	NBS	"	"
162	"	"	"	"	"	"	"	4787	9.4	"	"
163	"	"	"	"	"	"	"	4788	8.1	"	"
164	"	"	"	"	"	"	"	4791	8.8	"	"
165	"	UCLA 1093C	3520±80	"	2025 BC	"	"	4792	8.4	8.40	"
166	"	"	"	"	"	"	"	4793	8.4	"	"

No.	Source	Lab#	C14 Date	Material	Cal. Date	Reference	Assoc.	Lab#	Hydra	Mean	Analyst
167	Coso	UCLA 1093D	3580±80	Charcoal	2060 BC	Clewlow et al '70	Direct	4794	11.4	8.48	Bennett
168	"	"	"	"	"	"	"	4795	8.0	"	"
169	"	"	"	"	"	"	"	4796	8.2	"	"
170	"	"	"	"	"	"	"	4797	8.0	"	"
171	"	"	"	"	"	"	"	4798	8.3	"	"
172	"	"	"	"	"	"	"	4799	7.9	"	"
173	"	"	"	"	"	"	"	4800	8.4	"	"
174	"	"	"	"	"	"	"	4801	8.5	"	"
175	"	"	"	"	"	"	"	4802	7.9	"	"
176	"	"	"	"	"	"	"	4803	8.2	"	"
177	"	UCLA 1093E	3900±180	"	2390-2490BC	"	"	4804	8.4	8.16	"
178	"	"	"	"	"	"	"	4805	8.6	"	"
179	"	"	"	"	"	"	"	4806	8.1	"	"
180	"	"	"	"	"	"	"	4807	7.5	"	"
181	"	"	"	"	"	"	"	4808	8.2	"	"
182	"	UCLA 1093C	3520±80	"	2025 BC	"	Strat.	4809	8.1	8.00	"
183	"	"	"	"	"	"	"	4810	7.9	"	"
184	"	UCLA 1955	1910±60	"	80 AD	Berger & Libby n.d.	Direct	4890	11.8	12.27	"
185	"	"	"	"	"	"	"	4891	NBS	"	"
186	"	"	"	"	"	"	"	4892	12.9	"	"
187	"	"	"	"	"	"	"	4894	12.6	"	"
188	"	"	"	"	"	"	"	4895	11.7	"	"
189	"	"	"	"	"	"	"	4896	12.0	"	"
190	"	"	"	"	"	"	"	4897	11.6	"	"
191	"	"	"	"	"	"	"	4899	13.3	"	"
192	"	UCLA 1957	Modern	Collagen	-	"	"	4904	2.6	2.10	"
193	"	"	"	"	-	"	"	4905	1.8	"	"
194	"	"	"	"	-	"	"	4906	1.9	"	"
195	"	"	"	"	-	"	"	4907	NBS	"	"

No.	Source	Lab#	C14 Date	Material	Cal. Date	Reference	Assoc.	Lab#	Hydra NBS	Mean	Analyst
196	Coso	UCLA 1957	Modern	Collagen	-	Berger & Libby n.d.	Direct	4908	2.10	2.10	Bennett
197	"	UCLA 1789A	1440±50	Charcoal	360 AD	"	"	3342	4.4	4.62	"
198	"	"	"	"	"	"	"	3343	5.0	"	"
199	"	"	"	"	"	"	"	3345	4.3	"	"
200	"	"	"	"	"	"	"	3346	4.7	"	"
201	"	"	"	"	"	"	"	3347	4.7	"	"
202	"	UCLA 1789B	1960±50	"	30 AD	"	"	3348	4.4	4.60	"
203	"	"	"	"	"	"	"	3349	4.8	"	"
204	"	UCLA 1426A	810±60	Shell	1210 AD	"	"	3350	3.6	3.50	"
205	Modoc Glass Mt	none	none	-	-	-	-	54a	1.4	1.5	Johnson
206	"	"	"	"	"	"	"	54f	1.4	"	"
207	"	"	"	"	"	"	"	54i	1.7	"	"
208	"	Gak-1831	2180±80	Charcoal	180 BC	Johnson 1969	Direct	57f	2.8	2.45	"
209	"	"	"	"	"	"	"	58i	2.1	"	"
210	"	Gak-1832	2340±100	"	450 BC	"	"	59g	2.8	2.69	"
211	"	"	"	"	"	"	"	59i	2.5	"	"
212	"	"	"	"	"	"	"	59k	3.3	"	"
213	"	"	"	"	"	"	"	60e	2.8	"	"
214	"	"	"	"	"	"	"	60g	2.9	"	"
215	"	"	"	"	"	"	"	60h	3.0	"	"
216	"	"	"	"	"	"	"	60k	1.5	"	"
217	"	Gak-1833	2180±90	"	180 BC	"	"	62d	3.5	2.70	"
218	"	"	"	"	"	"	"	62h	1.9	"	"
219	"	Gak-1834	3470±80	"	1790-2010 BC	"	"	63b	3.1	3.50	"
220	"	"	"	"	"	"	"	63e	3.3	"	"
221	"	"	"	"	"	"	"	63f	3.4	"	"
222	"	"	"	"	"	"	"	63i	2.6	"	"
223	"	"	"	"	"	"	"	63j	3.2	"	"
224	"	"	"	"	"	"	"	64c	3.3	"	"

No.	Source	Lab#	C14 Date	Material	Cal. Date	Reference	Assoc.	Lab#	Hydra	Mean	Analyst
225	Modoc Glass Mt	Gak-1834	3470±80	Charcoal	1790-2010 BC	Johnson 1969	Direct	64e	5.6	3.50	Johnson
226	"	Gak-1835	3450±90	"	1710-1800 2010 BC	"	"	65a	4.1	3.70	"
227	"	"	"	"	"	"	"	65b	4.5	"	"
228	"	"	"	"	"	"	"	65c	4.3	"	"
229	"	"	"	"	"	"	"	65d	3.1	"	"
230	"	"	"	"	"	"	"	65e	3.6	"	"
231	"	"	"	"	"	"	"	65g	3.0	"	"
232	"	"	"	"	"	"	"	65i	2.5	"	"
233	"	"	"	"	"	"	"	65k	2.4	"	"
234	"	"	"	"	"	"	"	66a	4.6	"	"
235	"	"	"	"	"	"	"	66c	4.6	"	"
236	"	Gak-1835	3450±90	"	1710-1800 2010 BC	"	"	66g	4.2	"	"
237	"	"	"	"	"	"	"	66i	3.5	"	"
238	"	Gak-1836	4260±100	"	2970 BC	"	"	70b	3.9	3.64	"
239	"	"	"	"	"	"	"	70c	5.0	"	"
240	"	"	"	"	"	"	"	70e	3.9	"	"
241	"	"	"	"	"	"	"	70f	4.7	"	"
242	"	"	"	"	"	"	"	70g	3.2	"	"
243	"	"	"	"	"	"	"	70h	3.6	"	"
244	"	"	"	"	"	"	"	70i	4.1	"	"
245	"	"	"	"	"	"	"	70j	3.1	"	"
246	"	"	"	"	"	"	"	70k	3.5	"	"
247	"	"	"	"	"	"	"	71b	2.2	"	"
248	"	"	"	"	"	"	"	71c	3.0	"	"
249	"	"	"	"	"	"	"	71d	2.9	"	"
250	"	"	"	"	"	"	"	71g	4.2	"	"
251	"	Gak-1837	4750±110	"	3525-3660 BC	"	"	72a	4.1	4.20	"
252	"	"	"	"	"	"	"	72b	3.9	"	"
253	"	"	"	"	"	"	"	72c	4.5	"	"
254	"	"	"	"	"	"	"	72d	4.4	"	"

No.	Source	Lab#	C14 Date	Material	Cal. Date	Reference	Assoc.	Lab#	Hydra	Mean	Analyst
255	Modoc Glass Mt	Gak-1837	4750±110	Charcoal	3525-3660 BC	Johnson 1969	Direct	72f	4.2	4.20	Johnson
256	"	"	"	"	"	"	"	73b	4.1	"	"
257	"	"	"	"	"	"	"	73f	4.2	"	"
258	"	Gak-1833 Gak-1839	4030±90 4500±110	"	2970 BC	"	"	75a	4.0	4.58	"
259	"	"	"	"	"	"	"	75b	2.7	"	"
260	"	"	"	"	"	"	"	75c	5.2	"	"
261	"	"	"	"	"	"	"	75g	6.4	"	"
262	"	Gak-1840	5750±130	"	4600 BC	"	"	76a	4.4	4.22	"
263	"	"	"	"	"	"	"	76b	3.3	"	"
264	"	"	"	"	"	"	"	76c	3.5	"	"
265	"	"	"	"	"	"	"	76e	4.6	"	"
266	"	"	"	"	"	"	"	76h	4.2	"	"
267	"	"	"	"	"	"	"	76j	5.3	"	"
268	"	Gak-2424	6161±130	"	5125 BC	"	Loose	89b	4.6	4.43	"
269	"	"	"	"	"	"	"	90c	4.4	"	"
270	"	"	"	"	"	"	"	90d	4.3	"	"
271	"	Gak-1841	1540±100	"	450 AD	"	Direct	131b	2.1	2.35	"
272	"	"	"	"	"	"	"	131c	2.3	"	"
273	"	"	"	"	"	"	"	131f	2.1	"	"
274	"	"	"	"	"	"	"	131h	2.9	"	"
275	Mono Craters	UCLA 1920A	430±110	Charcoal	1440 AD	Berger & Libby n.d.	Direct	4415	6.5	5.98	Bennett
276	"	"	"	"	"	"	"	4418	4.7	"	"
277	"	"	"	"	"	"	"	4425	6.2	"	"
278	"	"	"	"	"	"	"	4427	6.5	"	"
279	"	UCLA 1920B	1000±300	"	1020 AD	"	"	4428	6.6	6.45	"
280	"	"	"	"	"	"	"	4430	6.5	"	"
281	"	"	"	"	"	"	"	4431	6.6	"	"
282	"	"	"	"	"	"	"	4432	6.2	"	"

No.	Source	Lab#	C14 Date	Material	Cal. Date	Reference	Assoc.	Lab#	Hydra	Mean	Analyst
283	Mono Craters	UCLA 1920B	1000±300	Charcoal	1020 AD	Berger & Libby n.d.	Direct	4433	6.3	6.45	Bennett
284	"				"	"	"	4434	6.5		"
285	"	UCLA 1093A	2040±145	"	130-370 BC	Clewlow et al '70	"	4778	8.4	8.40	"
286	"	UCLA 1093B	2900±80	"	1110-1210 BC	"	"	4790	7.7	7.70	"
287	Mono Glass Mt	UCLA 276	950±70	"	1025 AD	Fergusson & Libby '64	"	3076	3.4	4.23	"
288	"	"	"	"	"	"	"	3079	NBS	"	"
289	"	"	"	"	"	"	"	3083	4.2	"	"
290	"	"	"	"	"	"	"	3084	5.0	"	"
291	"	"	"	"	"	"	"	3085	7.1/4.3		"
292	"	UCLA 277	1580±80	"	360 AD	"	"	3088	4.1	4.40	"
293	"	"	"	"	"	"	"	3089	4.7		"
294	"	M-646	3080±300	Bone	1320-1425 BC	Crane & Griffin '59	Strat.	3145	5.7/6.1	5.90	"
295	"	Assoc.			100 BC	Bennyhoff per.comm.		3125	4.7	4.70	"
296	"	UCLA 1771B	2040±350	Collagen	130 BC	Berger & Libby n.d.	"		5.8	5.80	"
297	"	UCLA 1952A	1620±400	"	400 AD	"	Direct	4825	2.5	2.50	"
298	Mt. Konocti	UCLA 1795A	10260±340	"	-	Ericson & Berger '74	"	3757	4.6	4.60	"
299	"	"	"	"	-	"	"	3758	4.6	"	"
300	"	"	"	"	-	"	"	3760	4.6	"	"
301	"	UCLA 1795B	9040±210	"	-	Berger & Libby n.d.	"	4089	4.0	4.08	"
302	"	"	"	"	-	"	"	4090	3.8	"	"
303	"	"	"	"	-	"	"	4091	3.6	"	"
304	"	"	"	"	-	"	"	4092	NBS	"	"
305	"	"	"	"	-	"	"	4093	3.8	"	"
306	"	"	"	"	-	"	"	4094	4.3	"	"
307	"	"	"	"	-	"	"	4095	5.0	"	"

No.	Source	Lab#	C14 Date	Material	Cal. Date	Reference	Assoc.	Lab#	Hydra	Mean	Analyst
308	Mt. Konocti	UCLA 1795C	7750±400	Collagen	-	Berger & Libby n.d.	Direct	4096	3.8/4.7	4.40	Bennett
309	"	"	"	"	-	"	"	4097	4.9	"	"
310	"	"	"	"	-	"	"	4098	4.1	"	"
311	"	"	"	"	-	"	"	4099	5.2	"	"
312	"	"	"	"	-	"	"	4101	4.3	"	"
313	"	"	"	"	-	"	"	4102	3.6	"	"
314	"	M-647	4350±250	Bone	2900 BC	Crane & Griffin '58	Strat.	3132	3.1	3.10	"
315	"	"	"	"	"	"	"	3135	5.0	4.05	"
316	Napa Glass Mt				900 AD		Strat.		4.5	2.87	Unknown
317	"				"		"		4.1	"	"
318	"				"		"		2.8	"	"
319	"				"		"		2.4	"	"
320	"				"		"		2.7	"	"
321	"				"		"		1.9	"	"
322	"				"		"		3.7	"	"
323	"				"		"		2.2	"	"
324	"				"		"		4.2	"	"
325	"				"		"		2.3	"	"
326	"				"		"		2.5	"	"
327	"				"		"		3.3	"	"
328	"				"		"		2.4	"	"
329	"				"		"		2.2	"	"
330	"				"		"		2.8	"	"
331	"				"		"		-	"	"
332	"				"		"		3.2	"	"
333	"				"		"		2.3	"	"
334	"				"		"		2.2	"	"
335	"				1300 AD		"	3126	2.3	2.17	Bennett

No.	Source	Lab#	C^{14} Date	Material	Cal. Date	Reference	Assoc.	Lab#	Hydra	Mean	Analyst
336	Napa Glass Mt.				1300 AD		Strat.	3127	2.6	2.17	Bennett
337	"				"		"	3128	1.6		"
338	"	M-646	3080±300	Bone	1320-1425 BC	Crane & Griffin '59	Direct	3146	3.0	3.00	"
339	"	-	-	-	50 BC	"	Strat.	3123	3.4/12.9	3.40	"
340	"	-	-	-	"	"	"	3124	3.4		"
341	"	M-647	4350±250	Bone	2900 BC	Crane & Griffin '58	"	3129	2.6/NBS	2.60	"
342	"	M-886	1080±200	Wood	870 AD	Crane & Griffin '60	Direct	3140	0.6	1.59	"
343	"	"	"	"	"	"	"	3141	3.7		"
344	"	"	"	"	"	"	"	3142	1.2/NBS		"
345	"	"	"	"	"	"	"	3150	1.5		"
346	"	"	"	"	"	"	"	3151	1.7		"
347	"	"	"	"	"	"	"	3152	1.8		"
348	"	"	"	"	"	"	"	3153	1.3/NBS		"
349	"	M-884	500±150	"	1410 AD	"	"	3136	1.5	2.36	"
350	"	"	"	"	"	"	"	3137	1.7		"
351	"	"	"	"	"	"	"	3138	4.1		"
352	"	"	"	"	"	"	"	3139	1.5		"
353	"	"	"	"	"	"	"	3147	4.9		"
354	"	"	"	"	"	"	"	3148	1.6		"
355	"	"	"	"	"	"	"	3149	1.2		"
356	"	-	-	Collagen	900 AD		Strat.	3154	2.0	2.00	"
357	"	GX-659	2675±135	Collagen	880 BC	Ragir '72	Direct	3155	4.7	4.45	"
358	"	"	"	"	"	"	"	3156	4.2		"
359	"	UCLA 1794C	115±45	Charcoal	1680 AD	Berger & Libby n.d.		4104	2.1	2.30	"
360	"	"	"	"	"	"		4106	2.5		"
361	"	UCLA 1793A	440±50	"	1400 AD	"		3706	1.8	2.45	"
362	"	"	"	"	"	"		3707	2.6		"

No.	Source	Lab#	C14 Date	Material	Cal. Date	Reference	Assoc.	Lab#	Hydra	Mean	Analyst
363	Napa Glass Mt	UCLA 1793A	440±50	Charcoal	1400 AD	Berger & Libby n.d.	Direct	3708	2.0/4.4	2.45	Bennett
364	"	"	"	"	"	"	"	3709	1.8	"	"
365	"	"	"	"	"	"	"	3710	2.8	"	"
366	"	"	"	"	"	"	"	3711	2.2/4.6	"	"
367	"	"	"	"	"	"	"	3712	2.6/4.3	"	"
368	"	"	"	"	"	"	"	3713	2.7	"	"
369	"	"	"	"	"	"	"	3714	2.6	"	"
370	"	"	"	"	"	"	"	3715	2.7	"	"
371	"	"	"	"	"	"	"	3716	2.2	"	"
372	"	"	"	"	"	"	"	3717	3.4	"	"
373	"	UCLA 1793B	585±50	"	1380 AD	"	"	3718	2.6	2.22	"
374	"	"	"	"	"	"	"	3719	2.3	"	"
375	"	"	"	"	"	"	"	3721	1.7	"	"
376	"	"	"	"	"	"	"	3722	2.1	"	"
377	"	"	"	"	"	"	"	3723	2.6	"	"
378	"	"	"	"	"	"	"	3724	1.9	"	"
379	"	"	"	"	"	"	"	3725	1.7	"	"
380	"	"	"	"	"	"	"	3726	2.3	"	"
381	"	"	"	"	"	"	"	3727	2.6	"	"
382	"	"	"	"	"	"	"	3728	2.5	"	"
383	"	"	"	"	"	"	"	3729	2.1	"	"
384	"	UCLA 1793C	365±50	"	1460 AD	"	"	3730	2.2	2.36	"
385	"	"	"	"	"	"	"	3732	3.3	"	"
386	"	"	"	"	"	"	"	3733	2.2	"	"
387	"	"	"	"	"	"	"	3734	2.3	"	"
388	"	"	"	"	"	"	"	3736	2.3	"	"
389	"	"	"	"	"	"	"	3737	2.5	"	"
390	"	"	"	"	"	"	"	3738	1.8	"	"
391	"	"	"	"	"	"	"	3739	2.3	"	"
392	"	"	"	"	"	"	"	3740	2.1	"	"

No.	Source	Lab#	C14 Date	Material	Cal. Date	Reference	Assoc.	Lab#	Hydra	Mean	Analyst
393	Napa Glass Mt	UCLA 1793C	365±50	Charcoal	1460 AD	Berger & Libby n.d.	Direct	3741	2.6	2.36	Bennett
394	"	UCLA 1793D	465±50	"	1405 AD	"	"	3742	2.3	2.55	"
395	"	"		"	"	"	"	3744	2.0		"
396	"	"		"	"	"	"	3745	2.0	"	"
397	"	"		"	"	"	"	3746	3.0	"	"
398	"	"		"	"	"	"	3748	2.1	"	"
399	"	"		"	"	"	"	3749	2.1	"	"
400	"	"		"	"	"	"	3750	4.8	"	"
401	"	"		"	"	"	"	3751	2.8	"	"
402	"	"		"	"	"	"	3752	2.1	"	"
403	"	"		"	"	"	"	3753	2.3	"	"
404	"	UCLA 1891B	3270±70	Collagen	1625 BC	"	"	4039	3.8	3.90	"
405	"	"		"	"	"	"	4040	4.3		"
406	"	"		"	"	"	"	4042	4.7	"	"
407	"	"		"	"	"	"	4043	2.8	"	"
408	"	"		"	"	"	"	4044	3.0	3.0	"
409	"	UCLA 1891A	3050±130	"	1300 BC	"	"	4045	4.4	3.13	"
410	"	"		"	"	"	"	4047	1.8		"
411	"	"		"	"	"	"	4048	3.2	"	"
412 (W)	"	UCLA 1951A	1950±60	Charcoal	60 AD	"	"	4815	2.3	2.30	"
413	"	M-123	2880±300	"	1120-1200 BC	Crane '56	"	4819	3.8	"424"	"
414	"	UCLA 1953	345±50	"	1450 AD	Berger & Libby n.d.	"	4837	2.2	2.59	"
415	"	"		"	"	"	"	4841	2.0/4.1		"
416	"	"		"	"	"	"	4842	3.0		"
417 (S)	"	"		"	"	"	"	4833	2.7	"	"
418	"	"		"	"	"	"	4834	2.5	"	"
419	"	"		"	"	"	"	4839	3.7	"	"
420	"	"		"	"	"	"	4845	2.0	"	"
421(E Dago)	"	UCLA 1950	385±85	"	1500 AD	"	"	4811	NBS	2.60	"

No.	Source	Lab#	C14 Date	Material	Cal. Date	Reference	Assoc.	Lab#	Hydra	Mean	Analyst
422	Napa Glass Mt (E Dago)	UCLA 1950	385±85	Charcoal	1500 AD	Berger & Libby n.d.	Direct	4812	NBS	2.60	Bennett
423	"	"	"	"	"	"	"	4813	2.6		"
424	"	M-123	2880±300	"	1120-1200 BC	Crane '56	"	4820	4.1	3.87	"
425	"	"	"	"	"	"	"	4821	3.7		"
426	"	UCLA 1952B	1200±100	Collagen	700 AD	Berger & Libby n.d.	"	4829	2.7	2.70	"
427	"	UCLA 1953	345±50	Charcoal	1450 AD	"	"	4836	2.9	2.68	"
428	"	"	"	"	"	"	"	4835	2.3	3.30	"
429	"	"	"	"	"	"	"	4838	1.8	2.68	"
430	"	"	"	"	"	"	"	4840	2.8		"
431	"	"	"	"	"	"	"	4844	2.6		"
432	"	"	"	"	"	"	"	4846	2.2		"
433	"	"	"	"	"	"	"	4860	2.6		"
434	"	"	"	"	"	"	"	4862	2.7		"
435	"	"	"	"	"	"	"	4863	1.9		"
436	(W Dago)	UCLA 1958	2150±190	Collagen	170-360 BC	"	"	4909	3.3	-	-
437	(Hill 450)	UCLA 1953	345±50	Charcoal	1450 AD	"	"	4843	2.4		"
438	Obsidian Butte	UCLA 1815	870±80	"	1110-1210 AD	"	Strat.	3164	3.6	2.77	"
439	"	"	"	"	"	"	"	3165	2.6		"
440	"	"	"	"	"	"	"	3166	2.1/0.8		"
441	"	UCLA 1816	215±60	"	1670 AD	"	"	3163	2.7	2.70	"
442	Mt. Hicks	UCLA 276	950±70	"	1025 AD	Fergusson & Libby '64	Direct		14.4/5.4	5.40	"
443	"	UCLA 277	1580±80	"	360 AD	"	"		3.4	6.77	"
444	"	"	"	"	"	"	"		10.0		"
445	"	"	"	"	"	"	"		6.9		"
446	"	UCLA 1952A	1620±400	Collagen	400 AD	Berger & Libby n.d.	"		2.7	2.20	"
447	"	"	"	"	"	"	"		1.7		"

No.	Source	Lab#	C14 Date	Material	Cal. Date	Reference	Assoc.	Lab# Hydra	Mean	Analyst
448	Pine Grove Hills	UCLA 276	950±70	Charcoal	1025 AD	Fergusson & Libby '64	Direct	4.3	4.30	"
449	"	M-647	4350±250	Bone	2900 BC	Crane & Griffin '58	Strat.	4.3	4.80	"
450	"	"	"	"	"			5.3	"	"

BIBLIOGRAPHY

Abrams, D. n.d. Part 2, Salvage investigation at Little Pico Creek site. Unpublished manuscript 1968. UCLA Arch. Survey, Manuscript File 121.

Aiello, P. V. 1969. The chemical composition of rhyolitic obsidian and its effect on hydration rate: Some archaeological evidence. M.A. thesis. Department of Anthropology, UCLA.

Ambrose, W. 1976. Intrinsic hydration rate dating of obsidian. In Advances in Obsidian Glass Studies, edited by R. E. Taylor, pp. 81-105. Park Ridge, N.J.: Noyes Press.

Anderson, C. A. 1936. Volcanic history of the Clear Lake area, California. Geol. Soc. America Bull. 47(1):629-664.

Baedecker, P. B. 1976. SPECTRA: A computer program. In Advances in Obsidian Glass Studies, edited by R. E. Taylor, pp. 334-350. Park Ridge, N.J.: Noyes Press.

Baumhoff, M. A. 1955. Excavation of site Teh-1 (Kingley Cave). Univ. Calif. Arch. Survey, Report 40.

Baumhoff, M. A. 1957. An introduction to Yana archaeology. Univ. Calif. Arch. Survey, Report 41.

Baumhoff, M. A. 1963. Ecological determinants of aboriginal California population. Univ. Calif. Publ. in Amer. Arch. and Ethnology 49:231.

Baumhoff, M. A., and D. Olmsted 1963. Palaihnihan: Radiocarbon Support for Glottochronology. Amer. Anthro. 65:278-284.

Bean, L. J. 1971. An environmental interpretation of the social systems of some southern California aboriginal populations. Presentation at the Annual Meeting of the Society for California Archaeology, Sacramento.

Beardsley, R. K. 1948. Culture sequence in central California archaeology. Amer. Antiquity 14:1-29.

Beardsley, R. K. 1954. Temporal and areal relationships in central California, part I. Univ. Calif. Arch. Survey, Report 24:1-62.

Bell, R. E. 1977. Obsidian hydration studies in highland Ecuador. Amer. Antiquity 42(1):68-78.

Belshaw, C. S. 1965. Traditional Exchange and Modern Markets. Englewood Cliffs, N.J.: Prentice-Hall.

Bennyhoff, J. A. 1950. California fish spears and harpoons. Berkeley: Univ. Calif. Anthro. Records 9:295-337.

Bennyhoff, J. A. 1956. An appraisal of the archaeological resources of Yosemite National Park. Univ. Calif. Arch. Survey, Report 34:1.

Bennyhoff, J. A. 1958. The desert West: A trial correlation of culture and chronology. In Current views on Great Basin archaeology. Univ. Calif. Arch. Survey, Report 42:98-112.

Bennyhoff, J. A. 1977. Ethnography of the Plains Miwok. Center for Arch. Research at Davis, Publ. Univ. Calif. Davis 5, 181 pp.

Bennyhoff, J. A., and R. F. Heizer 1958. Cross-dating Great Basin sites by California shell beads. In Current views on Great Basin archaeology. Univ. Calif. Arch. Survey, Report 42:113-192.

Berger, R. 1970. The potential and limitations of radiocarbon dating in the Middle Ages: The radiochronologist's view. In Scientific Methods in Medieval Archaeology, ed. R. Berger, pp. 89-139. Berkeley: Univ. Calif. Press.

Berger, R., and J. E. Ericson 1974. Natural solid solutions: obsidians and tektites. In Volume 3, Recent Advances in Sciences and Technology of Materials, edited by A. Bishay, pp. 187-190. New York: Plenum Press.

Berger, R., and W. F. Libby n.d. UCLA radiocarbon dating list (in preparation).

Berry, B. J. L. 1967. Geography of market centers and retail distribution. Englewood Cliffs, N.J.: Prentice-Hall.

Bettinger, R. L., and T. F. King 1971. Interaction of political organization: a theoretical framework for archaeology in Owens Valley, California. UCLA Arch. Survey, Annual Report 13:139-150.

Bettinger, R., and R. E. Taylor 1974. Suggested revisions in interior southern California archaeological sequences. Univ. Nevada, Papers of Arch. Survey 5:4-26.

Binford, L. R. 1965. Archaeological systemics and study of culture process. Amer. Antiquity 31:203-221.

Binford, L. R. 1967. Invention and technology. Anthropology class lecture, April 13, 1967. Department of Anthropology, UCLA.

Bowman, H. R., F. Asaro, and I. Perlman 1973. On the uniformity of composition in obsidians and evidence for magmatic mixing. Jour. Geology 81:312-327.

Brand, D. D. 1938. Aboriginal trade routes for sea shells in the Southwest. Assoc. Pacific Coast Geographers 4:3-10.

Brown, A. K. 1967. The aboriginal population of the Santa Barbara Channel. Univ. Calif. Arch. Survey, Report 69:1-100.

Bruchner, R. 1965. Der Einfluss des Hydroxylgehaltes auf die Dichte und auf den Diffusions mechanismus in Kiesel-gläsern. Glastechnische Berichte 38:153-156.

Carron, J. P. 1966. La Conductiute electrique des obsidiennes entre 400°C et 950°C. Comptes Rendus Acad. Sci. 263:1665-1668.

Carron, J. P. 1969. Recherches sur la viscosite et les phenomenes de transport des ions alcalins dans les obsidiennes granitiques. Travaux du Laboratoire de Géologie, Ecole Normale Supérieure, Paris, No. 3, 112 pp.

Chagnon, N. 1970. Ecological and adaptive aspects of California shell money. UCLA Arch. Survey, Annual Report 1970:1-25.

Chang, J. 1957. Global distribution of the annual range in soil temperature: Trans. Amer. Geophys. Union 38(5)718-723.

Charles, R. J. 1958. Static fatigue of glass, I. Jour. Applied Phys. 29(11) 1549-1553.

Chase, P., and E. L. Davis n.d. Some obsidian hydration dates from Obsidian Butte, Imperial County, California. Unpublished manuscript, 1969.

Childress, J., and J. L. Chartkoff n.d. An archaeological survey of the English Ridge Reservoir in Lake and Mendocino Counties, California. Unpublished manuscript, 1966. UCLA Arch. Survey, Manuscript File MS-57:1.

Clark, D. L. 1961a. The application of the obsidian dating method to the archaeology of central California. Ph.D. dissertation, Stanford University.

Clark, D. L. 1961b. The obsidian dating method. Current Anthro. 2:111-116.

Clark, D. L. 1964. Archaeological chronology in California and the obsidian hydration method: part 1. UCLA Arch. Survey, Annual Report 1963-64: 141-209.

Clarke, D. L. 1968. Analytical Archaeology. London: Methuen.

Clewlow, C. W., R. F. Heizer, and R. Berger 1970. An assessment of radiocarbon dates for the Rose Spring site (CA-INY-372), Inyo County, California. Univ. Calif. Arch. Research Facility, Contributions 7:19-25.

Clifton, H. E. 1967. Paleographic significance of two Middle Miocene basalt flows, southeastern Caliente Range, California. USGS Prof. Paper 575-B: B32-B39.

Cobean, R. H., M. D. Coe, E. A. Perry, Jr., K. K. Turekian, and D. P. Kharkar 1971. Obsidian trade at San Lorenzo Tenochtitlan, Mexico. Science 174:666-671.

Colton, H. S. 1941. Prehistoric trade in the Southwest. Scientific Monthly 52:308-319.

Cook, S. F. 1943. The conflict between the California Indian and white civilization: I — The Indian versus the Spanish mission. Univ. Calif. Publ. Ibero-Americana 21:Appendix.

Cook, S. F. 1955a. The aboriginal population of the San Joaquin Valley, Calif. Univ. Calif. Anthro. Records 16:1.

Cook, S. F. 1955b. The aboriginal population of the north coast of California. Univ. Calif. Anthro. Records 16:127.

Cook, S. F., and R. F. Heizer 1962. Chemical and physical analysis of the Hotchkiss Site (CCo-138). Univ. Calif. Arch. Survey, Report 57:1-57.

Crane, H. R. 1956. University of Michigan Radiocarbon Dates I. Science 124(3224):664-672.

Crane, H. R., and J. B. Griffin 1958. University of Michigan Radiocarbon Dates III. Science 128(3332):1117.

Crane, H. R., and J. B. Griffin 1959. University of Michigan Radiocarbon Dates IV. Amer. Jour. Sci. Radiocarbon Suppl. 1;173-198.

Crane, H. R., and J. B. Griffin 1960. University of Michigan Dates V. Amer. Jour. Sci. Radiocarbon Suppl. 2:31-48.

Crank, J. 1956. The Mathematics of Diffusion. Oxford: Clarendon Press.

Curtis, F., and D. A. Frederickson 1964. Investigations at CCo-309, a protohistoric site in interior Contra Costa County, California. Unpublished manuscript.

Davis, J. T. 1961. Trade routes and economic exchange among the Indians of California. Univ. Calif. Arch. Survey, Report 54:1-71.

Davis, J. T. 1962. The Rustler Rockshelter Site (SBr-288), a culturally stratified site in the Mohave Desert, California. Univ. Calif. Arch. Survey, Report 57:27.

Davis, J. T., and A. E. Treganza 1959. The Patterson Mound: A comparative analysis of the archaeological site Ala-328. Univ. Calif. Arch. Survey, Report 47:1-92.

deAngulo, J., and L. S. Freeland 1929. Notes on the Northern Paiute of California. Jour. Soc. des Americantes de Paris 21:313-335.

DeGarmo, G. D. 1977. Identification of prehistoric intrasettlement exchange. In Exchange Systems in Prehistory, edited by T. K. Earle and J. E. Ericson, pp. 153-170. New York: Academic Press.

Dubois, C. 1935. Wintun ethnography. Univ. Calif. Publ. in American Arch. and Ethnology 36(1):1-148.

Dubois, C., and D. Demetracoupoulou 1931. Wintu myths. Univ. Calif. Publ. in American Arch. and Ethnology 28(5):279-403.

Earle, T. K. 1977a. A reappraisal of redistribution: complex Hawaiian chiefdoms. In Exchange Systems in Prehistory, edited by T. K. Earle and J. E. Ericson, pp. 213-229. New York: Academic Press.

Earle, T. K. 1977b. Exchange systems in archaeological perspective. In Exchange Systems in Prehistory, edited by T. K. Earle and J. E. Ericson, pp. 3-12. New York: Academic Press.

Earle, T. K. and J. E. Ericson (Editors) 1977. Exchange Systems in Prehistory. New York: Academic Press.

Elsasser, A. B. 1960. The archaeology of the Sierra Nevada in California and Nevada. Univ. Calif. Arch. Survey, Report 51.

Elsasser, A. B., and R. F. Heizer 1966. Excavation of two northwestern California coastal sites. Univ. Calif. Arch. Survey, Report 67:1-15.

Ericson, J. E. 1969. Obsidian hydration dating: studies on the chemical and physical parameters affecting the hydration of obsidian. Unpublished manuscript, Isotope Laboratory, Inst. Geophysics and Planetary Physics, UCLA.

Ericson, J. E. 1973a. On the archaeology, chemistry and physics of obsidian. Unpublished Master's Thesis, Department of Anthropology, UCLA.

Ericson, J. E. 1973b. Prehistoric trade in California—a preliminary study. Presentation of the 38th Annual Meeting of the Society for American Archaeology, San Francisco, California, May 3-5, 1973.

Ericson, J. E. 1975. New results in obsidian hydration dating. World Archaeology 7:151-159.

Ericson, J. E. n.d. Geoscience at Castaic Creek site, LAn-324. Unpublished manuscript, 1972, Isotope Laboratory, Institute of Geophys. and Planetary Physics, UCLA.

Ericson, J. E., and R. Berger 1974. Late Pleistocene American obsidian tools. Nature 249:824-825.

Ericson, J. E., and R. Berger 1976. Physics and chemistry of the hydration process in obsidians II: experiments and measurements. In Advances in Obsidian Glass Studies, edited by R. E. Taylor, pp. 46-62. Park Ridge, N.J.: Noyes.

Ericson, J. E., and R. Goldstein 1980. Work space: a new approach to the analysis of energy expenditure within site catchments. In Catchment Analysis: Essays on Prehistoric Resource Space, edited by F. J. Findlow and J. E. Ericson, pp. 21-30. Anthropology UCLA 10(1-2).

Ericson, J. E., and T. A. Hagan n.d. The establishment of an archaeological data bank for California. UCLA Arch. Survey, Manuscript File MS-1.

Ericson, J. E., T. A. Hagan, and C. W. Chesterman 1976. Prehistoric obsidian sources in California I: geological and geographical aspects. In Advances in Obsidian Glass Studies, edited by R. E. Taylor, pp. 218-239. Park Ridge, N.J.: Noyes.

Ericson, J. E., and Kimberlin, J. 1975. The chemical characterization of obsidian sources in California, Oregon, and Nevada by instrumental neutron activation analysis. Manuscript in preparation.

Ericson, J. E., J. D. MacKenzie, and R. Berger 1976. Physics and chemistry of the hydration process in obsidians I: theoretical implications. In Advances in Obsidian Glass Studies, edited by R. E. Taylor, pp. 25-45. Park Ridge, N.J.: Noyes.

Ericson, J. E., A. Makishima, J. D. MacKenzie, and R. Berger 1975. Chemical and physical properties of obsidian: a naturally occurring glass. Jour. Non-Crystalline Solids 17:129-142.

Ericson, J. E., and C. A. Singer 1977. Four basic hypotheses in lithic analysis. Proc. Conf. on Lithic Use-Wear, Vancouver, B.C.

Evans, R. K. n.d. The Monterey Peninsula College sites (Mnt-371, 372, 373)—preliminary report. Unpublished manuscript, 1967. UCLA Arch. Survey, Manuscript File MS-64:1.

Eyre, S. R. 1963. Vegetation and Soils. Chicago: Aldine.

Farmer, M. F. 1937. An obsidian quarry near Coso Hot Springs. Southwest Museum Masterkey 11:7-9.

Felton, E. L. 1965. California's Many Climates. Palo Alto, Calif: Pacific Book Publishers.

Fenenga, F., and F. Riddell 1949. Excavation of Tommy Tucker Cave, Lassen County, California. Amer. Antiquity 14:203-214.

Fergusson, G. J., and W. F. Libby 1964. UCLA Radiocarbon Plates III. Radiocarbon 6:318-339.

Findlow, F. J., V. C. Bennett, J. E. Ericson, and S. P. DeAtley 1975. A new obsidian hydration rate for certain obsidians in the American Southwest. Amer. Antiquity 40(3):345-348.

Fisher, H. T. 1963. SYMAP 1. Laboratory for Computer Graphics and Spatial Analysis, Harvard University, 1:1.

Fitzwater, R. J. 1962. Final report of Two Seasons excavation at El Portal, Mariposa County. Univ. Calif. Arch. Survey, Annual Report 1961-1962: 235-282.

Fitzwater, R. J. 1968. Big Oak Flat: two archaeological sites in Yosemite National Park. Univ. Calif. Arch. Survey, Annual Report 1968:275-318.

Fitzwater, R. J. n.d.-1. Excavations at Crane Flat, Yosemite National Park. Unpublished manuscript, 1964. UCLA Arch. Survey, Manuscript File MS-67:1.

Fitzwater, R. J. n.d.-2. The Big Oak Flat Report: the Hodgon Ranch site (Tuo-236). Unpublished manuscript, 1968. UCLA Arch. Survey, Manuscript File MS-131.

Fitzwater, R. J., and M. Van Vlissengen 1960. Preliminary report on an archaeological site at El Portal, California. UCLA Arch. Survey, Annual Report 2:155-178.

Flanagan, F. J. 1973. 1972 values of international geochemical reference samples. Geochem, Cosmochim. Acta 37:1189-1200.

Fowke, G. 1896. Stone art. Bur. Amer. Ethnology, Annual Report 13:57-257.

Frederickson, D. A. 1965. Recent excavations in the interior of Contra Costa County, California. Sacramento Anthro. Soc., Paper 3:18-25.

Frederickson, D. A. 1966. CCo-308: the archaeology of a middle horizon site in interior Contra Costa County, California. Unpublished M. A. thesis, Univ. Calif., Davis.

Frederickson, D. A. 1968. Archaeological investigation at CCo-30 near Alamo, Contra Costa County, California. UCD Center for Arch. Research Publication 1.

Frederickson, D. A. 1969. Technological change, population movement, environment adaptation, and the emergence of trade: inferences on culture change suggested by midden constituent analysis. UCLA Arch. Survey, Annual Report 11:105-125.

Frederickson, D. A. 1971. Changes in social ranking in a cultural sequence from Contra Costa County, California. Presentation at the Annual Meeting of the Society for California Archaeology, Sacramento.

Frederickson, D. A. 1973a. Early cultures of the north Coast Ranges, California. Unpublished Ph.D. dissertation, Department of Anthropology, Univ. Calif., Davis.

Frederickson, D. A. 1973b. Cultural diversity in early central California: a view from the north Coast Ranges. Presentation at Thirty-Eighth Annual Meeting of the Society for American Archaeology, San Francisco.

Friedman, I. 1968. Hydration rind dates of rhyolitic flows. Science 159(3817): 878.

Friedman, I., and W. D. Long 1976. Hydration rate of obsidian. Science 191:347-352.

Friedman, I., W. D. Long, and R. L. Smith 1963. Viscosity and water content of rhyolitic glass. Jour. Geophys. Res. 68:6523-6535.

Friedman, I., and N. Peterson 1971. Obsidian hydration dating applied to dating of basaltic volcanic activity. Science 172:1028-1029.

Friedman, I., K. L. Pierce, J. D. Obradovich, and W. D. Long 1973. Obsidian hydration dates glacial loading? Science 180(4087):733.

Friedman, I., and R. L. Smith 1960. A new dating method using obsidian, I: the development of the method. Amer. Antiquity 25(4):476-522.

Friedman, I., R. L. Smith, and D. L. Clark 1963. Obsidian dating. In Science in Archaeology, edited by D. Brothwell and E. Higgs, pp. 47-58. London: Praeger.

Friedman, I., R. L. Smith, and D. L. Clark 1969. Obsidian dating. In Science in Archaeology, edited by D. Brothwell and E. Higgs, pp. 62-73, 2nd Edition. London: Thames and Hudson.

Friedman, I., R. L. Smith, and W. D. Long 1966. The hydration of natural glass and the formation of perlite. Geol. Soc. America Bull. 77:323-330.

Garner, B. J. 1968. Models of urban geography and settlement location. In Socio-Economic Models in Geography, edited by R. J. Chorley and P. Haggett, pp. 303-360. London: Methuen.

Gebhardt, C. L. n.d. Archaeological salvage excavation of Sac-166 near Natoma, California. Manuscript, 1962. UCLA Arch. Survey, Manuscript File MS-82.

Geiss, J., and D. C. Hess, 1958. Argon-potassium ages and the isotopic composition of argon from meteorites. Astrophys. Jour. 127:224-236.

Gonsalves, W. C. 1955. Winslow Cave, a mortuary site in Calaveras County, California. Univ. Calif. Arch. Survey, Report 29:31.

Gordus, A. A., G. A. Wright, and J. B. Griffin 1968. Obsidian sources characterized by neutron activation analysis. Science 161:382-384.

Griffin, J. B., A. A. Gordus, and G. A. Wright 1969. Identification of the sources of Hopewellian obsidian in the Middle West. Amer. Antiquity 34:1-14.

Haekel, J., 1958. Zur Frage alter Kulturbeziehungen swischen Alaska, Kalifornien und dem Pueblo-Gebiet. Proc. 32nd International Congress of Americanists, 1956, Copenhagen, pp. 88-96.

Haggett, P. 1966. Locational Analysis in Human Geography. New York: St. Martin's.

Haller, W. 1963. Concentration-dependent diffusion coefficient of water in glass. Physics and Chemistry of Glasses 4(6):217-220.

Harner, M. J. 1958. Lowland Patayan phases in the lower Colorado River Valley and Colorado Desert, Univ. Calif. Arch. Survey, Report 42:93-97.

Harrington, M. R. 1948. An ancient site at Borax Lake, California. Southwest Museum Paper 16:1.

Harrison, W. M., and E. S. Harrison 1966. An archaeological sequence for the hunting people of Santa Barbara, California. UCLA Arch. Survey, Annual Report 8:17.

Heizer, R. F. 1941a. Aboriginal trade between the Southwest and California. Southwest Museum Masterkey 15:185-188.

Heizer, R. F. 1941b. The direct historical approach in California archaeology. Amer. Antiquity 7:98-122.

Heizer, R. F. 1946. The occurrence and significance of southwestern grooved axes in California. Amer. Antiquity 11:187-193.

Heizer, R. F. 1949. The archaeology of central California I. The early horizon. Univ. Calif. Anthro. Records 1:1.

Heizer, R. F. 1953. Archaeology of the Napa region. Univ. Calif. Anthro. Records 12(6):225-358.

Heizer, R. F. 1960. California population densities: 1770 and 1950. Univ. Calif. Arch. Survey, Report 50(79):9-11.

Heizer, R. F. 1964. The western coast of North America. In Prehistoric Man in the New World, edited by J. F. Jennings and E. Norbeck, pp. 117-148. Chicago: Univ. Chicago Press.

Heizer, R. F. 1974. Studying the Windmiller Culture. In <u>Archaeological Researches in Retrospect</u>, edited by G. R. Willey, pp. 177-204. Cambridge, Mass.: Winthrop.

Heizer, R. F., and A. B. Elsasser 1953. Some archaeological sites and cultures of the central Sierra Nevada. <u>Univ. Calif. Arch. Survey, Report</u> 21:1.

Heizer, R. F., and A. E. Treganza 1944. Mines and quarries of the Indians of California. <u>Calif. Div. Mines and Geology, Dept. State Mineralogist</u> 40(3):291-359.

Hodder, I. 1974. Regression analysis of some trade and marketing patterns. <u>World Archaeology</u> 6:172-189.

Horton, C. W., W. B. Hempkins, and A. A. J. Hoffman 1964. A statistical analysis of some aeromagnetic maps from the Northwestern Canadian Shield. <u>Geophysics</u> 29:582-601.

Horton, C. W., A. A. J. Hoffman, and W. B. Hempkins 1962. Mathematical analysis of the microstructure of an area at the bottom of Lake Travis. <u>Texas Jour. Sci.</u> 14:131-142.

Huggins, M. L. 1944. Density of glass. <u>Jour. Optical Soc. America</u> 34:420.

Hunt, A. 1960. Archaeology of the Death Valley salt pan. <u>Univ. Utah, Anthro. Paper</u> 47:1.

Irwin-Williams, C. 1977. A network model for the analysis of prehistoric trade. In <u>Exchange Systems in Prehistory</u>, edited by T. K. Earle and J. E. Ericson, pp. 141-150. N.Y.: Academic Press.

Jack, R. N. 1976. Prehistoric obsidian in California I: geochemical aspects. In <u>Advances in Obsidian Glass Studies,</u> edited by R. E. Taylor, pp. 183-217. Park Ridge, N.J.: Noyes.

Jack, R. N., and I. S. E. Carmichael 1969. The chemical "fingerprinting" of acid volcanic rocks. <u>Calif. Div. Mines and Geology, Short Contrib., Special Report</u> 100:17-30.

Jackson, D. 1975. Stepwise regression: BMDP2R. In <u>Biomedical Computer Programs</u>, edited by W. J. Dixon, pp. 491-439. Los Angeles: Univ. Calif. Press.

Jackson, D., and J. Douglas 1975. Multiple linear regression: BMDP1R. In <u>Biomedical Computer Programs</u>, edited by W. J. Dixon, pp. 453-489. Los Angeles: Univ. Calif. Press.

Jackson, D. D. 1972. Interpretation of inaccurate, insufficient, and inconsistent data. <u>Jour. Geophys. Res.</u> 28:97-109.

Jackson, T. L. 1974. The economics of obsidian in central California prehistory: applications of X-ray fluorescence spectrography in archaeology. Unpublished M.A. thesis, San Francisco State Univ.

Jaeger, J. C. 1961. The cooling of irregular-shaped igneous bodies. <u>Amer. Jour. Sci.</u> 259:721-734.

Johannsen, A. 1939. A Descriptive Petrography of the Igneous Rocks (2nd Edition). Chicago: Univ. Chicago Press.

Johnson, J. J. n.d. Archaeological investigation at 4-Sis-258. Manuscript File MS-115.

Johnson, LeRoy, Jr. 1969. Obsidian hydration rate for the Klamath Basin of California and Oregon. Science 165(3900):1354-1355.

Katsui, Y., and Y. Kondo 1965. Dating of stone implements by using the hydration layer in obsidian. Japanese Jour. Geol. Geog. 46(2-4):45-60.

Kerr, P. F. 1959. Optical Mineralogy (3rd Edition). New York: McGraw-Hill.

Kimberlin, J. 1971. Obsidian chemistry and the hydration dating technique. Unpublished M.S. Thesis, Department of Anthropology, UCLA.

Kimberlin, J. 1976. Obsidian hydration rate determinations of chemically characterized samples. In Advances in Obsidian Glass Studies, edited by R. E. Taylor, pp. 63-80. Park Ridge, N.J.: Noyes.

King, C. T. 1967. The Sweetwater Mesa site (LAn-267) and its place in southern California prehistory. UCLA Arch. Survey, Annual Report 9: 25-76.

King, C. T. 1971. Chumash inter-village economic exchange. The Indian Historian 4(1):30-43.

KIng, C. T. 1973. An explanation of differences and similarities of beads. Presentation of Thirty-eighth Annual Meeting, Society for American Archaeology, San Francisco.

King, C. T., T. Blackburn, and E. Chandonet 1968. The archaeological investigation of three sites on the Century Ranch, western Los Angeles County, California. UCLA Arch. Survey, Annual Report 10:1.

King, T. 1970. The dead at Tiburon. Northwestern Calif. Arch. Soc., Occasional Paper 2:1.

King, T. 1972. New views of California Indian societies. The Indian Historian 5(4):12-17.

Klimek, S. 1935. Culture element distributions: I, the structure of California Indian culture. Univ. Calif. Publ. in American Arch. and Ethnology 37 (1):1-70.

Kniffen, F. B. 1926. Achomawi geography. Univ. Calif. Publ. in American Arch. and Ethnology 23(5):297-332.

Kowta, M. n.d. Preliminary report on an excavation in Kern County, California. Manuscript, 1954. UCLA Arch. Survey, Manuscript File MS-29.

Kowta, M., and J. C. Hurst 1960. Site Ven-15: the Truinfo rockshelter. UCLA Arch. Survey, Annual Report 2:201-217.

Kroeber, A. L. 1923. The history of native culture in California. Univ. Calif. Publ. in American Arch. and Ethnology 20:125-142.

Kroeber, A. L. 1925. Handbook of the Indians of California. Bureau of American Ethnology, Bulletin 78.

Kroeber, A. L. 1962. Two papers on the aboriginal ethnology of California. Univ. Calif. Arch. Survey, Report 56.

Krumbein, W. C., and F. A. Graybill 1965. An Introduction to Statistical Models in Geology. New York: McGraw-Hill.

Lankford, P. 1974. A guide to SUPERMAP. Classnotes, UCLA.

Lanning, P. 1963. Archaeology of the Rose Spring site, Iny-372. Univ. Calif. Publ. in American Arch. and Ethnology 49(3):237-336.

Lathrap, D. W., and D. Shutler, Jr. 1953. An archaeological site in the High Sierra of California. Amer. Antiquity 20:226.

Layton, T. N. 1973. Temporal ordering of surface-collected obsidian artifacts by hydration measurement. Archaeometry 15:129-132.

Lee, R. 1969. Chemical temperature integration. Jour. Applied Meteorology 8:423-430.

Lee, R. R., D. A. Leich, T. A. Tombrello, J. E. Ericson, and I. Friedman 1974. Obsidian hydration profile measurements using a nuclear reaction technique. Nature 250:44-47.

Leich, D. A., and T. A. Tombrello 1973. A technique for measuring hydrogen concentration versus depth in solid samples. Nuclear Instruments and Methods 108:67-71.

Leich, D. A., T. A. Tombrello, and D. S. Burnett 1973. The depth distribution of hydrogen in lunar materials. Earth Planetary Sci. Lett. 19:305-314.

Lillard, J. B., R. F. Heizer, and F. Fenenga 1939. An introduction to the archaeology of central California. Sacramento Jr. College, Department of Anthropology, Bulletin 2.

Loeb, E. M. 1936. Pomo folkways. Univ. Calif. Publ. in American Arch. and Ethnology 19(2):152-205.

Marshall, R. R. 1961. Devitrification of natural glass. Geol. Soc. America Bull. 72(10):1493-1520.

Mauss, M. 1954. The Gift. London: Cohen and West.

McNitt, J. R. 1968. Geology of the Kelseyville quadrange, Sonoma, Lake, and Mendocino Counties. Calif. Div. Mines and Geology, Map Sheet No. 9.

Meighan, C. W. 1953a. Preliminary excavation at the Thomas site, Martin County. Univ. Calif. Arch. Survey, Report 19:1.

Meighan, C. W. 1953b. The Coville rockshelter, Inyo County, California. Univ. Calif. Anthro. Records 12(5):171.

Meighan, C. W. 1955a. Notes on the archaeology of Mono County, California. Univ. Calif. Arch. Survey, Rpeort 28:6.

Meighan, C. W. 1955b. Excavation of Isabella Meadows Cave, Monterey County. Univ. Calif. Arch. Survey, Report 29:1.

Meighan, C. W. 1955c. Archaeology of the north Coast Ranges, California. Univ. Calif. Arch. Survey, Report 30:1-39.

Meighan, C. W. 1970. Obsidian hydration rates. Science 170(3953):99-100.

Meighan, C. W. 1974. The archaeology of Amapa, Nayarit, Mexico. Manuscript, 1974.

Meighan, C. W., and V. C. Bennett 1978. Obsidian dates II: a compendium of theobsidian determination made at the UCLA Obsidian Hydration Dating Laboratory. Inst. Archaeology, Arch. Survey, UCLA (in press).

Meighan, C. W., L. J. Foote, and P. V. Aiello 1968. Obsidian dating in west Mexican archaeology. Science 160(3832):1069-1075.

Meighan, C. W., and C. Vance Haynes 1970. The Borax Lake site, revisited. Science 167:1213-1221.

Michels, J. W. 1965. Lithic serial chronology through obsidian hydration dating. Unpublished Ph.D. dissertation, Department of Anthropology, UCLA.

Michels, J. W. 1967. Archaeology and dating by hydration of obsidian. Science 158(3798):211-214.

Michels, J. W. 1973. Dating Methods in Archaeology. N.Y.: Seminar Press.

Michels, J. W. n.d. Archaeological investigation of the Mammouth Junction site (Mno-382). Manuscript, 1964. UCLA Arch. Survey, Manuscript File MS-59.

Michels, J. W., and C. Bebrich 1971. Obsidian hydration dating. In Dating Techniques for the Archaeologist, edited by H. N. Michaels and E. K. Ralph, pp. 164. Cambridge, Mass.: MIT Press.

Moratto, M. J. 1972. A study of the prehistory in the southern Sierra Nevada foothills. Ph.D. dissertation, Department of Anthropology, Univ. Oregon (Xerox Univ. Microfilms, Inc.).

Morgenstein, M., and T. J. Riley 1973. Hydration-rind dating of basaltic glass: a new method for archaeological chronologies. Asian Perspectives 17:145.

Morgenstein, M., and T. J. Riley 1975. Hydration-rind dating of basaltic glass: a new method for archaeological chronologies. Manuscript, Univ. Hawaii.

Morgenstein, M., and P. Rosendahl 1976. Basaltic glass hydration dating in Hawaiian archaeology. In Advances in Obsidian Glass Studies, edited by R. E. Taylor, pp. 141-164. Park Ridge, N.J.: Noyes.

Moriarity, J. R., G. Shumway, and C. N. Warren 1959. Scripps estate site I (SDi-525). UCLA Arch. Survey, Annual Report 1:187-216.

Muffler, L. J. P., and D. E. White 1969. Active metamorphism of upper Cenozoic sediments in the Salton Sea geothermal field and the Salton Trough, southeastern California. Geol. Soc. America Bull. 80(2):157-182.

Nasedkin, V. V. 1964. Volatile components of volcanic glasses. Geochem. International 2:317-335.

Newcomer, M. H. 1971. Some quantitative experiments in handaxe manufacture. World Archaeology 3(1):85-94.

O'Connell, J. F. 1968. Elko eared/Elso corner-notched projectile points as time markers in the Great Basin. Univ. Calif. Arch. Survey Report 70:129-140.

O'Connell, J. F., and R. D. Ambro 1968. A preliminary report on the archaeology of the Rodriguez site (CA-LAS-194), Lassen County, California. Univ. Calif. Arch. Survey, Report 73:95-194.

Odum, E. P., and H. T. Odum 1959. Fundamentals of Ecology (2nd Edition). Philadelphia: W. B. Saunders.

O'Keefe, J. A. 1964. Water in tektite glass. Jour. Geophys. Res. 69:3701-3707.

Parks, G. A., and T. T. Tieh 1966. Identifying the geographical source of artifact obsidian. Nature 211(5046):289-290.

Pendergast, D. M., and C. W. Meighan 1959. The Greasy Creek site, Tulare County, California. UCLA Arch. Survey, Annual Report 1:1-10.

Perlman, I., F. Asaro, and H. V. Michel 1972. Nuclear applications in art and archaeology. Annual Rev. Nuclear Sci 22:383-426.

Pincus, H. J., and M. B. Dobrin 1966. Geological application of optical data processing. Jour. Geophys. Res. 24:42-61.

Polanyi, K. 1957. The economy as instituted process. In Trade and Market in Early Empires, edited by K. Polanyi, C. M. Arensberg, and H. W. Pearson, N.Y.: The Free Press.

Pope, S. T. 1918. Yahi archery. Univ. Calif. Publ. in American Arch. and Ethnology 13(3):103-152.

Price, J. A. 1967. Conditions in the development of silent trade. Kroeber Anthro. Assoc. 36:67-79.

Ragir, S. 1972. The early horizon in central California prehistory. Univ. Calif. Arch. Res. Facility, Contributions 15:1-329.

Ragir, S., and J. Lancaster 1966. Analysis of a surface collection from High Rock Canyon, Nevada. Univ. Calif. Arch. Survey, Report 66:1-35.

Rappaport, R. A. 1968. Pigs for the Ancestors, Ritual in the Ecology of a New Guinea People. New Haven: Yale Univ. Press.

Rasson, J. 1966. Excavation at Ahwahee, Yosemite National Park. UCLA Arch. Survey, Annual Report 8:165-183.

Renfrew, C., 1969. Trade and culture process in European prehistory. Current Anthro. 10(2-3):151-169.

Renfrew, C., 1977. Alternative models for exchange and spatial distribution. In Exchange Systems in Prehistory, edited by T. K. Earle and J. E. Ericson, pp. 71-90. N.Y.: Academic Press.

Renfrew, C., J. E. Dixon, and J. R. Cann 1968. Further analysis of Near Eastern obsidians. Proc. of the Prehistoric Society, Series 2(34):319.

Riddell, F. A. 1951. The archaeology of a Paiute village site. Univ. Calif. Arch. Survey, Report 12:14-28.

Riddell, F. A. 1956. Final report on the archaeology of Tommy Tucker Cave. Univ. Calif. Arch. Survey, Report 35:1-25.

Riddell, F. A. 1960. The archaeology of the Karlo site (Las-7), California. Univ. Calif. Arch. Survey, Report 53:1.

Rogers, D. B. 1929. Prehistoric Man of the Santa Barbara Coast. Santa Barbara, Calif.: Santa Barbara Museum of Natural History.

Rogers, D. B. 1939. Early lithic industries of the lower basin of the Colorado River and adjacent desert areas. San Diego Museum, Paper 3:1.

Rogers, D. B. 1945. An outline of Yuman prehistory. Southwestern Jour. Anthro. 1:167-198.

Ruby, J. 1966. Archaeological investigation of Big Tujunga site. UCLA Arch. Survey, Annual Report 8:91-150.

Sahlins, M. D. 1965. On the sociology of primitive exchange. In Relevance Of Models for Social Anthropology, edited by M. Banton, pp. 139-236. London: ASA, Monographs.

Sahlins, M. D. 1972. Stone Age Economics. Chicago: Aldine.

Sanders, W. T. 1956. The central Mexican symbiotic region: a study in prehistoric settlement patterns. In Prehistoric Settlement Patterns in the New World, edited by G. R. Wiley. Viking Fund Publications in Anthropology 23:115-127.

Scholze, H. 1959. Der Einbau des Wassers in Gläsern. Glastechnische Berichte 32:81-88, 142-152, 278-281, 314-320.

Scholze, H. 1966. Gases and water in glass: Part II. Lectures on Glass Science and Technology. N.Y.: Glass.

Shinem, J. A. 1972. Field Guide to Landforms in the United States. N.Y.: Macmillan.

Sidrys, R. 1977. Mass-distance measures for the Maya obsidian trade. In Exchange Systems in Prehistory, edited by T. K. Earle and J. E. Ericson, pp. 91-105. N.Y.: Academic Press.

Singer, C. A., and J. E. Ericson 1977. Quarry analysis at Bodie Hills, Mono County, California: a case study. In Exchange Systems in Prehistory, edited by T. K. Earle and J. E. Ericson, pp. 171-188. N.Y.: Academic Press.

Smith, C. E., and W. D. Weymouth 1952. Archaeology of the Shasta Dam area, California. Univ. Calif. Arch. Survey, Report 18:1-36.

Smith, E. I. 1973. Mono Craters, California: a new interpretation of the eruptive sequence. Geol. Soc. America Bull. 84:2685-2690.

Smith, J. E., and J. M. LaFaue n.d. Excavation of site SLO-297, Vaqueros Reservoir, San Luis Obispo County, California. Manuscript 1959-60. UCLA Arch. Survey, Manuscript File MS-116.

Sterud, J. 1965. Archaeological investigations at the Mammouth Junction site: second season. UCLA Arch. Survey, Manuscript File.

Stevenson, D. P., F. H. Stross, and R. F. Heizer 1971. An evaluation of X-ray fluorescence analysis as a method for correlating obsidian artifacts with source location. Archaeometry 13(1):17-25.

Steward, J. H. 1933. Ethnography of the Owens Valley Paiute. Univ. Calif. Publ. in American Arch. and Ethnology 32:233-350.

Steward, J. H. 1938. Basin-plateau aboriginal sociopolitical groups. Bur. Amer. Ethnology Bull. 70:1-346.

Suess, H. E. 1970. Bristlecone-pine calibration of the radiocarbon time scale 5200 B.C. to the present. In Radiocarbon variations and absolute chronology, edited by I. V. Olson, pp. 303-308. Stockholm: Almquist and Wiksell.

Suzuki, M. 1973. Potential of obsidian hydration dating. Jour. Faculty of Sci., Univ. Tokyo, Section V, 4:241.

Swartz, C. A. 1954. Some geometrical properties of residual maps. Geophysics 19:46-70.

Taylor, W. W. 1961. Archaeology and language in western North America. Amer. Antiquity 27:71-81.

Tourtellot, G., and J. A. Sabloff 1972. Exchange systems among the ancient Maya. Amer. Antiquity 37:126-135.

Tower, D. B. 1945. The use of marine mollusks and their value in reconstructing prehistoric trade routes in the American Southwest. Papers of the Excavator's Club, Cambridge, Mass., 2:3.

Townsend, J. B. 1960. Two rockshelters and a village site in Borrengo State Park. UCLA Arch. Survey, Annual Report 2:249-276.

Treganza, A. E. 1954. Salvage archaeology in Nimbus and Redbank Reservoir areas, central California. Univ. Calif. Arch. Survey, Report 26:1.

Treganza, A. E., and M. H. Heicksen n.d. Salvage archaeology in the Black Butte Reservoir area, Glenn County, California. Manuscript, 1961. UCLA Arch. Survey, Manuscript File MS-21.

Treganza, A. E., and C. G. Malamud 1950. The Topanga culture, first season's excavation of the Tank site, 1947. Univ. Calif. Arch. Survey, Report 12(4):129.

True, D. L., E. L. Sterud, and E. L. Davis 1967. An archaeological survey at Indian Ranch, Panamint Valley, California. UCLA Arch. Survey, Annual Report 9:1-24.

Tsuboi, C. 1959. Application of double fourier series on computing gravity anomalies and other gravimetrical quantities at higher elevations from surface gravity anomalies. Ohio State Univ., Inst. Geol., Photogrammetry, and Cartography, Report 2:1.

U.S. Department of Commerce 1954. Plane coordinate intersection tables (2½ minute) for California. U.S. Dept. Commerce, Special Publ. 327: 1-75. Washington, D.C.: U.S. Govt. Printing Office.

Valastro, S., Jr., E. M. Davis, and A. G. Varela 1975. University of Texas at Austin Radiocarbon Dates X. Radiocarbon 17(1):63.

Van de Geer, J. P. 1971. Introduction to Multivariate Analysis for the Social Sciences. San Francisco: W. H. Freeman.

Vayda, A. P. 1967. Pomo trade feasts. In Tribal and Peasant Economies, edited by G. Dalton, pp. 494-500. N.Y.: Natural History Press.

Voeglin, E. W. 1942. Culture elements distribution, XX, northeast California. Univ. Calif. Anthro. Records 7(2):47.

Von Neumann, J., and O. Morgenstein 1955. Theory of Games and Economic Behavior (3rd Ed.). Princeton, N.J.: Princeton Univ. Press.

Von Werlhof, J. n.d.-1. Archaeological investigation at Tul-145 (Cobble Lodge). Manuscript, 1959-1961. UCLA Arch. Survey, Manuscript File MS-137.

Von Werlhof, J. n.d.-2. Archaeological investigation at Hospital Rock, Tulare County, California. Manuscript, 1960. UCLA Arch. Survey, Manuscript File MS-136.

Wallace, W. J. 1951. The archaeological deposits of Moaning Cave, Calaveras County. Univ. Calif. Arch. Survey, Report 12:229.

Wallace, W. J. 1955. Suggested chronology for southern California coastal archaeology. Southwestern Jour. Anthro. 11:214-230.

Wallace, W. J. 1960. Archaeological resources of Buena Vista watershed, San Diego County, California. UCLA Arch. Survey, Annual Report 2: 277-306.

Wallace, W. J. 1962. Prehistoric cultural developments in the southern California deserts. Amer. Antiquity 28:172-180.

Wallace, W. J., and E. S. Taylor 1952. Excavation of Sis-13, a rockshelter in Siskiyou County, California. Univ. Calif. Arch. Survey, Report 15:13.

Warren, C. N. 1968. Cultural tradition and ecological adaptation of the southern California coast. Eastern New Mexico Univ., Contr. in Anthro. 1(3):1-14.

Warren, C. N. 1973. California. In The Development of North American Archaeology, edited by J. E. Fitting, pp. 231-249. Garden City, N.Y.: Anchor Books.

Webb, M. C. 1975. Exchange networks in prehistory. California, Palo Alto: Reviews in Anthro., pp. 357-383.

Wedel, W. R. 1941. Archaeological investigation at Buena Vista Lake, Kern County, California. Bur. Amer. Ethnology Bull. 131.

Weide, M. L. n.d. Excavation at Ora-82. Manuscript, 1967. UCLA Arch. Survey, Manuscript File MS-72.

Wilke, P. J. n.d. Archaeology of Peppertree site, area 2. Univ. Calif., Riverside. (Manuscript in preparation.)

Wilmsen, E. N. 1972. Social exchange and interaction. Museum of Anthro., Univ. Michigan, Anthro. Papers 46:1-17.

Wire, V. V. M. n.d. Alamo Creek site, San Luis Obispo County, California. Manuscript 1959-1961. UCLA Arch. Survey, Manuscript File MS-116.

Wright, G. A. 1969. Obsidian analyses and prehistoric Near Eastern trade: 7500-3500 B.C. Museum of Anthro., Univ. Mich., Anthro. Papers 37.

Wright, G. A. 1974. Archaeology and Trade. Addison-Wesley Module in Anthropology 49:1-48.

Wright, H., and M. Zeder 1977. The simulation of a linear exchange systems under equilibrium conditions. In Exchange Systems in Prehistory, edited by T. K. Earle and J. E. Ericson, pp. 233-254. N.Y.: Academic Press.

Zipf, G. K. 1949. Human behavior and the principle of least effort: an introduction to human ecology. Cambridge, Mass.: Addison-Wesley.

www.ingramcontent.com/pod-product-compliance
Ingram Content Group UK Ltd.
Pitfield, Milton Keynes, MK11 3LW, UK
UKHW060200240426
12048UKWH00029B/1675